More praise for
DOCTORS BY NATURE

"*Doctors by Nature* brings a new WOW on every page! This fascinating book will blow your mind with the chemical complexity of plants, the insight, ingenuity, and instincts of animals, and the healing powers within the reach of a thrilling array of creatures."

—SY MONTGOMERY, author of *The Soul of an Octopus: A Surprising Exploration into the Wonder of Consciousness*

"As Jaap de Roode reveals in *Doctors by Nature*, many animals seek out substances to relieve illnesses or battle parasites that drag their health down. . . . They could even teach us a thing or two about medicine."

—SAM KEAN, *Wall Street Journal*

"From ants to apes, bees to bears, and caterpillars to capuchins, animals use nature's pharmacy to recover from injury and treat and prevent disease. But how do animals know what plants and substances their bodies need? De Roode dives into the expanding, fascinating research surrounding the world of zoopharmacognosy—animal self-medication—illuminating the interwoven intricacies and innate brilliance of the animal and plant kingdoms, and our crucial role in protecting both."

—KAREN SHAW BECKER, coauthor of the *New York Times* bestseller *The Forever Dog Life*

"Jaap de Roode takes the reader through a breezy and fun exploration into the complex world of animal medication."
—ED CARA, *Gizmodo*

"In *Doctors by Nature*, biologist Jaap de Roode tells a witty tale of how animals use drugs to fight infections and stay alive. The book teaches us to see animals as capable creatures that look after their own health, proposing that animals' medicinal knowledge can inspire new approaches to animal welfare, agriculture, and even human medicine."
—BETH SHAPIRO, author of *How to Clone a Mammoth*

"A fascinating look at the medical knowledge of animals; this would be a worthy addition to any life sciences collection."
—*Library Journal*

"Outstanding! Jaap de Roode's engrossing investigation of the remarkable relationship between the animal and plant kingdoms reveals fascinating secrets of self-healing behavior by insects, birds, reptiles, and mammals. *Doctors by Nature* is replete with new and surprising biological discoveries, recounted with charm and enthusiasm."
—ADRIENNE MAYOR, author of *Gods and Robots: Myths, Machines, and Ancient Dreams of Technology*

"From chimpanzees treating parasites with bitter plants to monarch butterflies protecting their offspring with toxic milkweed, *Doctors by Nature* uncovers how animals instinctively use nature's pharmacy to heal themselves and their communities. Jaap de Roode weaves together cutting-edge research, engaging stories, and his own scientific journey to show how animal medication behaviors can inspire new approaches to human and veterinary medicine. This eye-opening book will forever change how you view the intelligence and ingenuity of our fellow creatures, while offering hope for new discoveries that could revolutionize health care for all species."

—NEIL SHUBIN, author of *Your Inner Fish*

"*Doctors by Nature* presents a scientifically grounded account of the fascinating strategies animals use to treat and prevent wounds and illnesses. Through masterful scientific storytelling, de Roode shares many remarkable examples of evolved behavioral adaptations that reveal our species to be only one of many that can heal themselves."

—BARBARA NATTERSON-HOROWITZ, MD,
coauthor of *Zoobiquity* and *Wildhood*

"[An] exceptional book investigating the myriad ways in which diverse animals across taxa self-medicate. . . . The book is written in a very readable style and is about storytelling as much as intellectual enlightenment."

—DAVID M. GASCOIGNE, *Travels with Birds*

DOCTORS BY NATURE

DOCTORS BY NATURE

How Ants, Apes, and Other Animals Heal Themselves

JAAP DE ROODE

PRINCETON UNIVERSITY PRESS
PRINCETON & OXFORD

Published by Princeton University Press
41 William Street, Princeton, New Jersey 08540
99 Banbury Road, Oxford OX2 6JX

press.princeton.edu

GPSR Authorized Representative: Easy Access System Europe - Mustamäe tee 50, 10621 Tallinn, Estonia, gpsr.requests@easproject.com

First paperback printing, 2026
Paperback ISBN 9780691239354

Library of Congress Cataloging-in-Publication Data for the cloth edition:

Names: De Roode, Jaap, 1977– author.
Title: Doctors by nature : how ants, apes, and other animals heal themselves / Jaap de Roode.
Description: Princeton : Princeton University Press, 2025. | Includes bibliographical references and index.
Identifiers: LCCN 2024022700 (print) | LCCN 2024022701 (ebook) | ISBN 9780691239248 (hardback) | ISBN 9780691239255 (ebook)
Subjects: LCSH: Health behavior in animals. | Animals—Drug use.
Classification: LCC QL756.6 .D47 2025 (print) | LCC QL756.6 (ebook) | DDC 591.5—dc23/eng/20240826
LC record available at https://lccn.loc.gov/2024022700
LC ebook record available at https://lccn.loc.gov/2024022701

British Library Cataloging-in-Publication Data is available

Editorial: Alison Kalett and Hallie Schaeffer
Production Editorial: Ali Parrington
Text Design: Katie Osborne
Cover Design: Ben Higgins
Production: Danielle Amatucci
Publicity: Matthew Taylor and Kate Farquhar-Thomson
Copyeditor: Susan Matheson

Cover image: Life on white / Alamy

This book has been composed in Arno Pro with Gamay

In memory of Jane Goodall and Frans de Waal, who did so much to show us that humans are not that different from other animals

CONTENTS

DISCLAIMER

The information presented in this book is not meant to be medical advice. Readers should consult a physician or a veterinarian for matters concerning their own health, and the health of their pets and other companion animals.

DOCTORS BY NATURE

1

Birds, Bees, and Butterflies

"Did you know that monarch butterflies use drugs?" I ask.

It is October 2022, and I am sitting at an outdoor table at the St. Marks National Wildlife Refuge located on the southern side of Florida's Big Bend, right on the water of the Gulf of Mexico. Behind me is a white lighthouse. To my left I can look out over Goose Creek Bay, where I saw dolphins earlier. In front of me is a lagoon with alligators. Standing around the table are twenty people, varying in age from about three years old to seventy years old. We are at the St. Marks Monarch Butterfly Festival, held every year on the fourth Saturday of October. It is one of my favorite weekends of the year. For more than a decade, my students and I have made the six-hour drive from Emory University in Atlanta, where I am a professor, to attend the celebration of monarch butterflies. We come here to tell people about monarchs, and the research we do on them.

"Just like us, monarchs get germs," I say to a little girl wearing a tutu and dress-up butterfly wings. Holding the butterfly for her to see, I explain that these germs make the monarchs very sick. "They cannot go to the doctor like you or me, but luckily they can find medicines in the plants they eat."

Most people who know about monarchs are aware of their amazing migration.[1] As temperatures drop and the days get shorter, monarch butterflies from the United States and Canada embark on an amazing autumn journey. Flying as many as forty-five hundred kilometers, they travel to oyamel fir forests in the Transvere Neovolcanic Belt in Central Mexico. From late October to late November, and coinciding with the Mexican celebration of Day of the Dead, hundreds of millions of monarchs make their way into the high-altitude forests. They form clusters on the trees. Although each monarch weighs about as little as a medium-sized paperclip, some branches break under the weight of the thousands of monarchs that huddle together. Monarchs overwinter at these sites until February and March, when they mate and fly back north. On their way south, many monarchs follow the coast of the Gulf of Mexico to reach their overwintering sites. And lots of these monarchs stop in the refuge at St. Marks. On a good day, we see thousands flying through the refuge, feasting on the nectar of the abundant salt bushes and other flowering plants.

Amazing as monarch migration is, though, it is not the reason I started studying monarchs. I like to tell people that I study monarchs because they get sick. This takes many people by surprise. Accustomed as we are to becoming sick ourselves, and taking our pets to the vet, few of us think of wild animals becoming ill. But they do. Just as humans encounter a whole collection of disease-causing viruses, bacteria, worms, and protozoans throughout our lifetime, so do birds, bees, and butterflies. The most common disease in monarchs is caused by a one-celled parasite called *Ophryocystis elektroscirrha*.[2] Because its name is difficult to pronounce, many people refer to the parasite simply as "OE." This parasite is somewhat related to the parasites that cause malaria in humans, and it is no joke for monarchs. The parasite forms millions of spores on the skin of the monarch and pokes little holes in the monarch body. If the parasite does not kill the monarch, it causes dehydration and weight loss. Infected monarchs cannot fly well. So, instead of completing their journey to the overwintering sites in Mexico, they die along the way.[3]

Sitting at the table in St. Marks, I show people how we figure out whether monarchs are infected. My students and I like to call it the monarch health check. Like nurses and doctors, we wear examination gloves as we stick a clear plastic sticker to the abdomen of the butterfly (it does not hurt them), then place the sticker on a paper index card. We use a microscope and check for little black parasite spores. I show the festival visitors the parasites when we find them.

Then I tell them something truly remarkable. *Monarch butterflies are expert doctors.* Just as humans use drugs to treat parasitic infections, so do monarchs. As it turns out, I tell my audience, when faced by this horrendous OE parasite, monarchs are not helpless. They can seek out medicinal plants that reduce infection and relieve disease symptoms.

POISONOUS PLANTS

I started studying the parasites of monarchs in 2005 when I moved to the United States for a research position. While I was initially interested in studying the basic biology of these parasites, I quickly became interested in the interactions between the parasites, the monarchs, and the caterpillars' sole food source, milkweeds. Like many other butterflies, monarchs are specialist herbivores, meaning they eat only specific plants as caterpillars. For monarchs, their specialized diet consists only of milkweeds. There are actually many kinds of milkweed, in many forms and sizes, but most are in the same genus, *Asclepias*. When we rip off their leaves, they ooze white latex that looks like milk—hence the name.

In addition to producing latex, milkweeds produce a class of chemicals known as cardenolides. These steroid chemicals are toxic to most animals, and the plants use them to deter herbivores.[4] Monarchs can tolerate them, though. What's more, the caterpillars, while feeding on the plants, store the toxic chemicals in their own tissues.[5] This is what makes monarchs poisonous to their predators. Monarchs have bright orange wings, lined with black lines and white spots, which they use to tell birds and other predators they taste bad.[6]

When I started studying monarchs, it was a well-known fact that monarchs use cardenolides to protect themselves against predators. But with my interest in parasites, a question soon started forming in my brain. I knew of studies that had shown that other types of toxic chemicals, found in other plants, can kill disease-causing viruses of insects.[7] And that made my colleagues and me wonder: Could the cardenolides found in milkweeds be toxic to OE parasites? Were the monarchs using plants not just as food, but as medicine?

To answer that question, I set up an experiment with two groups of monarch caterpillars—one group fed on only tropical milkweed and the other fed on only swamp milkweed. All caterpillars (a total of 240) were exposed to OE by feeding them milkweed with parasite spores. I knew from published studies that tropical milkweed (*Asclepias curassavica*) has more cardenolides than swamp milkweed (*Asclepias incarnata*). After the caterpillars became butterflies, we tested how many of the monarchs had become infected and how sick they were. If cardenolides could protect against parasites, we would expect the monarchs who had fed on tropical milkweed as caterpillars to experience less illness. The result was exciting: in the group of monarchs that had fed on tropical milkweed as caterpillars, 20 percent fewer monarchs became infected than those that had fed on swamp milkweed. And the tropical milkweed–fed caterpillars that did become infected had less than half the number of parasites and were a lot less sick, living up to twice as long.[8] All in all, our experiment suggested that highly toxic milkweed not only wards off predators but also acts as a potent antiparasitic drug.

The next question was as logical as it was unlikely. Was it possible that monarchs could intentionally take advantage of these medicinal milkweeds? Would sick monarchs be able to specifically use highly toxic milkweeds as a form of medicine? In 2008, I had taken a position as assistant professor at Emory University. There, my team and I carried out a series of experiments in which we offered infected and uninfected female butterflies medicinal tropical milkweed and nonmedicinal swamp milkweed in big flight cages. We counted the numbers of eggs that these females laid on each species. What we found is that infected butterflies laid way more eggs on medicinal plants than on nonmedicinal plants.[9] Uninfected butterflies did not. In other words: when monarch mothers are infected, they prefer to lay their eggs on medicinal milkweed.

That infected monarchs prefer to lay their eggs on medicinal plants is remarkable. It is even more remarkable when we think about who they are actually protecting. Diseased monarchs do not benefit from the medicinal plants themselves. They are already infected and have suffered the consequences. They cannot cure themselves. What they also cannot do is avoid the spread of parasites to their offspring. The parasites form

FIGURE 1.1. A female monarch butterfly lays eggs on a medicinal milkweed, which will reduce infection and disease symptoms in her offspring caterpillars. Photo by Jaap de Roode.

millions of spores on the butterflies' abdomens, and every time the butterfly lays an egg on a milkweed plant, some of these will inevitably get stuck to her eggs and the milkweed leaves.[10] But what the mother butterfly *can* do is choose to lay her eggs on medicinal plants. When her babies hatch from their eggs, they will ingest the parasites. But they will also ingest the medicinal milkweed. And this reduces the chance that the parasites take hold. Should the caterpillar still become infected, the plant reduces the growth of the parasites, and thereby relieves disease symptoms. Thus, rather than medicating herself, a monarch mother medicates her offspring—and she does so even before those offspring are born. A wonderful case of "mother knows best."

TEENY-TINY BRAINS

As I was studying monarch medication, I realized that many other animals use drugs as protection against disease. (I fully realize that humans are animals. But for the sake of simplicity, I will use the term "animals" specifically for nonhuman animals throughout this book). In the 1980s, primatologists had discovered that chimpanzees can use the toxins and hairy leaves of plants to treat intestinal worm infections. I found other studies that showed that goats and sheep are their own medical doctors too. And while many scientists traditionally believed that animals needed big brains to be able to medicate themselves—a bias mostly driven by the fact that chimpanzees are our closest living relatives—this idea did not jibe with the data.

I learned that woolly bear caterpillars and wood ants can use medicine too. Thus, animals with brains smaller than a pinhead can be just as good at medicating as those with brains like our own. What this suggested to me is that animal medication is common across the animal kingdom. I became fascinated with this idea, and that fascination eventually grew into this book. In the coming chapters, we'll explore all of these examples, and more.

Over the last four decades, scientists have discovered that animals do in fact seem to seek out medication (though, as we'll see, defining exactly what this means is a tricky task): animals of all shapes and sizes use a vast array of plants, fungi, toxic animals, chemicals, and other natural products to fight infections and alleviate disease. And they can do so in four different ways. First, animals can use "prophylaxis," which is when healthy animals choose to eat foods and antiparasitic compounds *before* they get sick to stay healthy and prevent disease. Japanese monkeys that live in areas with more parasites eat more antiparasitic plants than monkeys that live in areas with fewer parasites.[11] In Ethiopia, baboons that are at greater risk of schistosome infection increase their worm resistance by eating more toxic berries.[12] Second, animals may use "therapeutic medication." This is the use of medicinal compounds when the animal is already infected: chimpanzees suck the toxic juice out of bitter plants when sick with parasitic worms, and woolly bear caterpillars use toxic alkaloids to kill parasitic fly maggots. A third form of medication is "body anointing," where animals as diverse as lemurs, cats, and coatis rub antiparasitic substances into their fur to deter parasites such as mites, lice, and mosquitoes. Finally, animals may use "fumigation," by which they add antiparasitic substances to their living or sleeping quarters. Fumigation is used widely by birds, who line their nests with aromatic plants that kill mites, ticks, and lice. It is also used by ants and bees, which fill their nests with antimicrobial tree secretions to prevent disease.

Some scientists have described all these different behaviors with the word "zoopharmacognosy."[13] The term derives from the root words "zoo" (animal), "pharma" (drug), and "gnosy" (knowing). Other scientists prefer to describe the different behaviors as "animal self-medication." I do not particularly like either term. To me, the word "zoopharmacognosy" suggests that animals know that they are medicating themselves (they may not: the behaviors they display could be fully innate, as we will see in chapter 8). And the term "animal self-medication" suggests that animals exclusively medicate themselves (they do not—as we already saw, monarch butterflies medicate their offspring). I believe a more inclusive term to describe all these different behaviors is "animal medication," and that is what I will use throughout this book.

Demonstrating that animals use medication is difficult. I will discuss this in the next few chapters, highlighting the use of observational studies and experiments. For now, I want to make two important points. First, I will spend most of this book discussing behaviors that allow animals to fight against infection—that is, dealing with parasites and pathogens that make them sick. The reason for this focus is not only because parasites and pathogens are extremely important for animal evolution but also because most of the well-described examples of animal medication involve defenses against infections. That said, infectious diseases are not the only reason animals use medication. As this research continues, we are learning that animals may use medicine to treat wounds or relieve sore joints.[14] Orangutans, for example, mix specific plants with saliva and rub the mixture either into wounds,[15] or onto different body parts, which reduces inflammation.[16] Scientists have also suggested that pregnant animals may use particular plants to induce labor. Pregnant and lactating sifaka females increase consumption of tannin-rich plants, which is associated with increased body weight and stimulation of milk secretion.[17]

There are also reports of animals using drugs for recreational use. The concept of drunken monkeys is quite popular indeed. And reports of drunken elephants ransacking buildings can count on a wide readership. After a herd of 50 elephants raided a shop that sold a fermented drink in 2012, local police spokesman Asish Samanat commented in an online news outlet: "They were like any other drunk—aggressive and unreasonable but much, much bigger."[18] (Although it is entertaining to assume that animals consume alcohol to get drunk, a recent analysis suggests that primates actually eat fermented fruits because the fermentation process breaks down toxic chemicals that would otherwise make the fruits inedible, and because it provides a food source during times when fresh fruit is unavailable.[19]) Wound treatments, pregnancy care, and alcohol use fall outside the scope of this book, but if you are interested in learning more about these fascinating behaviors, I recommend the excellent book *Wild Health* by Cindy Engel.[20]

The second point I want to make is that it is not always obvious that a specific animal behavior is a form of medication. As I will discuss in the next few chapters, the most important criterion for animal medication is that the behavior helps the animal: either by reducing or avoiding infection, or by alleviating disease. But infection can also change behaviors in ways that are not beneficial to animals. On the contrary: it turns out that parasites and pathogens are masters at manipulating animal behaviors for their own benefit.[21] This means that when we see an infected animal change its behavior, we cannot simply assume it is medicating. Instead, it may be the parasites and pathogens doing the talking.

That animals can use medicine is now supported by many studies, and I chose to write this book to share the many fascinating stories of how animals use medication to protect themselves against parasites and pathogens. I also wanted to introduce the people who have provided us with these stories, showcasing how scientists from many different backgrounds, and in different continents, have been driven by an innate interest in understanding the natural world. As we will see, those scientists share a common belief: the study of animals is not only interesting in itself but can also teach us about how we can tackle the many parasites and pathogens that continue to inflict death and suffering on humans, domesticated animals, and pets.

The goal of this book is therefore twofold. First, I want to make the case that animals are highly evolved experts at medicine. They may not speak Latin, be trained in bedside manners, or write prescriptions, but they have evolved a plethora of ways to keep infectious diseases at bay through rather sophisticated medicine.

The second goal of this book is to show that we can benefit from studying animal doctors. Yes, most researchers studying animal medication are driven by a basic curiosity to understand the natural world. But as we will see, we can use insights into the medication behaviors of goats and sheep to increase animal health and reduce antibiotic resistance at farms and in the livestock industry. We can apply the antiparasitic behaviors of bees to improve beekeeping. And work is underway to develop bug repellents from compounds discovered by cats. Many scientists studying animal medication believe that their research may ultimately lead to the discovery of drugs that we can use to treat our own diseases.

Some would argue that our modern chemistry and technology equip us well enough to come up with new drugs from scratch.[22] But consider this: over the last forty years, more than half of the new antibacterial drugs and 45 percent of the antiparasitic drugs that hit the market for human use were derived from natural products. These include compounds from plants, bacteria, and fungi.[23] With the ongoing threat of infectious disease, and the ever-growing number of pathogens that evolve resistance to the drugs we rely on, it is more important than ever to study animal medication—and to apply the medical knowledge of animals to the health of humans and domesticated animals.

We'll explore all these ideas in future chapters. But before we get to those exciting applications, let's start by asking how a sick chimpanzee sparked the birth of a new scientific field.

2

Chimp Chausiku

As part of his address to the McGill Medical School in 1894, Canadian physician William Osler said: "The desire to take medicine is one feature which distinguishes man, the animal, from his fellow creatures."[1] Osler was one of the founders of Johns Hopkins Hospital, and he introduced the medical student residency program, wherein students learn at the bedside as much as in the classroom. Osler's remarks would have resonated well with his nineteenth-century Canadian audience, and they remain appealing today.

Indeed, for those of us living in contemporary Western societies, it seems important to find traits that make us unique, setting us apart from other animals.[2] For some time, scientists believed that humans were the only species that makes tools. We are not: Caledonian crows are master toolmakers, and many other animals, including chimpanzees, use tools too.[3] Some have argued that what makes us unique is our use of language. But many animals have some form of language too. The Japanese tit (a species of bird) uses a call that warns neighbors about predators and a recruitment call that tells others to come to a good resource. Amazingly, these calls even have syntax: the two calls are combined into a single call that tells other birds to come and help attack a predator.[4] Humans are also not unique in possessing culture. Many animals, from chimpanzees to bower birds, develop local customs and habits. Japanese monkeys living on Kōjima (Koshima) Island, for example, are famous for washing sweet potatoes in river water, a behavior that only happens in this and a few other populations.[5]

Osler was right that medical students learn from seeing and speaking to their patients, but he was wrong about his insistence that humans are the only species practicing medicine. Interestingly, the idea that only humans know how to use medicine is a modern, Western concept, driven by a desire to distance ourselves from nature. To a large extent, this desire stems from the influence of Christianity, the major religion in Western societies that is built on the importance of dualism: good versus evil, and human versus nature.[6] And so, as people in the Western world transitioned from a species that was connected to nature, and survived by hunting and gathering, to a species that believes in its own uniqueness, and buys food and medicine in fluorescent-lit mega supermarkets and pharmacies, we started to forget that we are one with nature. And as we forgot that, we also forgot that much of our knowledge used to stem from observing our fellow creatures.

Perhaps some of the best evidence of humans learning from animals comes from the knowledge obtained from bears. Much of traditional human medicine has its origins in shamanism, the religious practice that enables a person (shaman) to interact with what they believe to be the spirit world. Shamanism is a practice shared by many Indigenous cultures around the world. Shamans enter a different state of consciousness, and often rely on animal helpers that give strength, send dreams, and accompany the shaman on journeys into the spirit world. In the Southern Hemisphere, shamans often have lions, panthers, and tigers as their companions. In the Northern Hemisphere, the helper would often be a bear.

Following the habits of the bear, shamans would enter caves, sing sacred songs, and fast to become bears themselves. They would wear bear masks, bear skins, and necklaces or amulets made of bear teeth or claws. Some shamans would smoke a pipe in the shape of a bear, eat herbs similar to those that bears eat, and use bear gall, fat, and claws in healing ceremonies. Many would mimic bears hibernating. In their trance, shamans would experience the death and birth symbolized by bears entering their caves in winter and reemerging from them in spring. Thus, instead of distancing themselves from animals, shamans would try to become one with them to gain wisdom.[7]

The importance of bears for shamanism has been especially well-documented for Native American cultures.[8] For the Ojibwa (Chippewa) Midewiwin, the use of healing herbs was handed down from generation to generation; as members progressed in society, it was said they "followed the bear path."[9]

As Ojibwa medicine man Siyaka explains: "We consider the bear as chief of all animals in regard to herb medicine, and therefore it is understood that if a man dreams of a bear he will become an expert in the use of herbs for curing illness."[10]

Sioux medicine man Lame Deer similarly said: "The bear is the wisest of animals as far as medicines are concerned. If a man dreams of this animal, he could become a great healer."[11]

In a Cherokee story on the origin of disease, plants come together in a meeting to discuss what disease they will be able to cure. Sending a dream to a medicine man, they declared, "We, the green folk, will help you with any disease. But we are shy, you have to come to us and ask us for help when you are sick. You can also ask the bear because he best knows our qualities."[12]

Traditional healers would not only learn from bears—and other animals—through trance-like states but also by observing the behavior of bears directly. The Cheyenne observed bears eating yarrow, and they made tea out of yarrow leaves to treat colds. The Crow observed bears eating the leaves, stems, berries, and roots of kinnikinnick (also known as bearberry), and they applied powdered leaves to sores. The Navajo observed Kodiak bears dig up roots of a plant called Oshá (*Ligusticum porteri*), chew the roots, and then spit a mix of the root and saliva into their paws before thoroughly rubbing their fur.[13] At least fifteen different tribes use the plant, often referred to as bear root, to treat viral and bacterial infections.[14] These days, one can order Oshá root extract online.

While you may not have tried any of these bear medicines yourself, you probably have taken an aspirin at some point in your life. And as it turns out, the chemical on which this drug is based was most likely first discovered by bears as well. The hibernation of bears is a remarkable feat. Using dens that they burrow, or existing caves or hollows underneath trees, bears in North America, Europe, and Asia do not eat, drink, urinate, or defecate for a period of five to seven months. Their body temperature falls five degrees centigrade, and their heart rate slows from over forty beats per minute to less than ten. They can lose over a third of their body weight.[15] And then, as spring arrives, they wake up. As bears leave their dens, they start eating willow bark, willow buds, and meadowsweet shoots that contain salicylic acid. This acid is a chemical compound that was altered by chemists to produce aspirin. Bears use this natural aspirin to flush excess uric acid out of their blood and tissues, and to reduce back pain caused by laying still for so long. As Diarmuid Jeffreys writes in his book *Aspirin: The Remarkable Story of a Wonder Drug*, the discovery of aspirin and other medicines "could have started with mimicry; born from the observation

FIGURE 2.1. A Native American bear shaman. Painting by George Catlin, *Medicine Man, Performing His Mysteries over a Dying Man*, 1832, oil on canvas, 29 × 24 in. (73.7 × 60.9 cm), Smithsonian American Art Museum, Gift of Mrs. Joseph Harrison, Jr.

that a sick animal will seek out and eat a particular plant, or roll in a particular patch of grass. One can imagine how useful such knowledge would be to a hunter on the trail of a weak prey, and it could have been remembered and copied when the hunter himself fell ill."[16]

As we explore the science of animal medication in this book, it is important to remember that many of the scientific discoveries of the last decades are in fact rediscoveries of traditional knowledge. And as such, it is fitting to describe, in the next section, how a collaboration between a primatologist and a traditional healer resulted in what is widely considered the first convincing scientific evidence for the use of medicine by an animal.

BITTER LEAVES

"Someday I will travel to Africa to live with chimpanzees."

That's what primatologist Michael A. Huffman of Kyoto University told his mother at the age of four.[17] Inspired by the stories of the cheeky chimpanzee in *Curious George*, which his mother read to him as a bedtime routine, Huffman fell in love with the animals.[18]

"In high school I was Monkey Mike. I had chimpanzees on my brain. I was reading books, subscribing to *Scientific American* magazine. I was trying to get to Africa."

Years later, after he had met his first chimpanzee in the Denver Zoo as an Explorer Scout intern in high school, he enrolled in a study abroad program in Japan as a college freshman. Huffman then entered graduate school and completed his primatology training at the prestigious Kyoto University, the birthplace of primatology in Japan.

At the time, the focus of primatology in Japan was very different than in the United States and Europe.[19] Burdened by the dualistic focus instilled by cultural Christianity, primatologists in the United States and Europe did not view primates as individuals but rather as impersonal representatives of their species. They had to be studied as objectively as possible, without any opportunity for ascribing personalities or individual behaviors, lest the primatologists be accused of anthropomorphism.

Primatologists in Japan, however, were not burdened with the cultural need to distance themselves from nature. With their Buddhist and Shinto heritage, Japanese primatologists allowed themselves to view primates as individuals with personalities and feelings.[20] Where Western primatologists would give their study objects numbers, Japanese researchers would give them names. They would track their family relationships over time, and try to view the world from the primate's perspective.[21]

"It is the empathy that I felt toward animals that attracted me to Japan," says Huffman.

And as Huffman completed his training in the Japanese tradition, he would document the first scientific example of an animal medicating itself—thereby kick-starting a whole new scientific discipline.

It was November 21, 1987. Huffman had traveled to Mahale Mountains National Park in Tanzania to study the role of old individuals in chimpanzee society. Together with Mohamedi Seifu Kalunde, a senior game officer of Tanzania National Parks, he was tracking a small group of chimpanzees with two old females. The group also included Chausiku, a female approximately twenty-nine years old. Huffman had met her before, when he arrived in Mahale for the first time, two years earlier. Exhausted, dehydrated, and hungry from a long trip by boat down Lake Tanganyika, he was recuperating, sprawled out on a bench at the research base camp. Chausiku walked into the camp with her infant and stared at him. She was the first chimp he ever met in the wild.

Following the group of chimpanzees during the morning of that November day, Huffman and Mohamedi saw Chausiku climb a tree and build a nest.

"This is not the time chimpanzees typically do that," says Huffman. "I see her infant Chopin swinging around up in the tree out of her reach and trying to get in trouble. Usually, mom is right there, and looks after him, but for some reason she is not paying attention today."

The other members of the group retrieve the infant and take him with them as they move off to forage elsewhere, to return later. After a long rest, Chausiku leaves the tree, is reunited with Chopin, and then stops at a shrub that Huffman had not seen chimpanzees feed on before. She removes a branch and eats it in a peculiar way, peeling off the bark and leaves, and chewing and sucking on the pith (the inner part of the branch). She swallows the juice of the pith, but spits out the fibrous parts.

Mohamedi tells Huffman that this plant is known by the WaTongwe, the local people of which he was a senior member, as mjonso. Known by scientists as *Vernonia amygdalina*, its common name is bitter leaf. As the name implies, it is bitter, and Mohamedi explains that the WaTongwe use the leaves of the plant as medicine to treat a whole range of ailments, including stomach upset, intestinal parasites, malaria, and diarrhea. At this point a new idea nestles itself in Huffman's brain: Could it be that Chausiku was using the plant as medicine as well?

Huffman and Mohamedi decide not to continue following the old chimps but to track Chausiku instead. "All my colleagues said chimps don't get sick," says Huffman. "They said chimps are always healthy. But all the signs started to make sense to me. Making the bed, spending time in the bed. She pooped while laying sideways on a fallen tree. I had never seen that before. She could not walk very long without stopping to rest. She could apparently not stand up long enough to properly defecate." She also had no appetite and produced dark urine. All in all, Chausiku was clearly sick. That night, Huffman could not sleep.

He was too eager to learn more about Chausiku's behavior and the possible outcome of that bitter "medicine" she took.

The next morning, he and Mohamedi regrouped and found Chausiku. She was still moving slowly, but after the small group's noon nap, she suddenly stood up and took off with Chopin. Huffman and Mohamedi had to run to keep up with her. It was remarkable! According to Mohamedi, when the WaTongwe use the plant as medicine, recovery typically happens in twenty hours. That was exactly the time between Chausiku's chewing of the bitter pith and her apparent recovery the following day. "Those two days really convinced me that mjonso was responsible."

But convincing himself was not enough. Huffman needed more evidence to persuade his colleagues that Chausiku was truly using the plant as medicine. As a first step, Huffman started field tests to look at the toxic effects of the plant. He made watery solutions of the crushed leaves in what soup and rice bowls he could gather from their base camp. He added minnows, little fish, from a stream nearby. "Every two hours for 24 hours, I would check to see how they were doing. Some of the fish in the high doses died. That gave me the idea to collect leaves and take them back to Japan for chemical analysis."

On his return to Japan from Tanzania, Huffman handed over a big trash bag full of dried leaves to his Kyoto University colleagues Hajime Ohigashi and Koichi Koshimizu, experts in plant chemistry. They were more than happy to analyze the leaves. In fact, they had been waiting for an opportunity to analyze *Vernonia amygdalina*—mjonso—for some time, but the batches they had received from Cameroon were moldy and unusable by the time they arrived in Japan. Together with their students, they immediately started extracting the chemicals from Huffman's leaves, and found a number of sesquiterpene lactones, a class of chemicals well-known for their toxicity and medicinal value against parasitic worms, amoebae, bacteria, and even cancerous tumors.[22] They also found other compounds previously unreported in the scientific literature. They determined the chemical structures and activities of thirteen novel chemicals and named them steroid glucosides. The story that started unfolding was that Chausiku fell ill, used a plant with antiparasitic compounds, and recovered within twenty hours.

But was it really medication? Was Chausiku consuming it because she was sick? Could she have the knowledge that eating that particular plant would make her recover? *There is no way!* That's what most scientists thought.

"All my colleagues at Mahale said chimpanzees just eat this plant; it has nothing to do with medication; chimps never get sick." Huffman disagreed. He saw all the signs of illness. But he needed to find out what made them sick. "One doctor of tropical medicine at the Pasteur Institute back in Kyoto said: try looking at parasites. So, I learned about parasite life cycles. I started to look at parasite loads. How many chimps have parasites? How do infections change over the season? I took fecal samples for detecting parasites and urine for other signs of infection. I used little plastic fish-shaped containers that are normally used as lunchbox sushi soy sauce containers to suck up the urine off leaves and rocks. I analyzed chimpanzee activity, and I started to see more and more sickness. I put together daily medical charts for twenty individuals and followed them over a couple of years. But I needed another observation of bitter pith chewing. I needed to know if mjonso affected parasites."

In 1991, Huffman got that next observation.[23] He and Mohamedi were once again studying chimps in Mahale. They had their eyes on Fatuma, a twenty-eight-year-old female, with an infant named Pim. They noted that Fatuma suffered from flatulence and yellow runny diarrhea. As Huffman had observed with Chausiku several years earlier, Fatuma was lethargic and spent a lot of her time resting. On December 23, Fatuma sucked the bitter juice from the pith of two shoots of *Vernonia.*

By the next day, Fatuma started recovering; she no longer had diarrhea, and she improved so much that she was able to hunt down a red colobus monkey. This was twenty-three hours after she swallowed the bitter pith juice. When analyzing stool samples from Fatuma collected on December 23 and 24, Huffman recorded a large drop in the number of eggs of a parasitic worm, known as *Oesophagostomum stephanostomum*, that makes its living in the gut. Back in the lab, Ohigashi and Koshimizu later calculated that the amount of a chemical named vernonioside B1, which Fatuma ingested by sucking the pith, was roughly the same amount that WaTongwe healers give to their human patients. Overall, then, Huffman demonstrated that a wild chimpanzee ingested an effective dose of a medicinal plant that coincided with a reduction in parasitic worms and recovery from disease.

Huffman carried out another study from 1989 to 1992. Collecting dung samples from almost fifty chimpanzees, he found that chimpanzees were much more likely to be infected by *O. stephanostomum* in the wet season than in the dry season.[24] The worm is a nasty parasite and appears to account for most of the severe chimpanzee parasite infections in Mahale. Infected chimpanzees suffer from abdominal pain, bowel irritation, diarrhea, and general malaise. In one male chimpanzee the worms even spread from the gut to the urinary tract, killing him in the process. The fact that the worms were more common during the wet season makes sense, as chimpanzees pass worm eggs in their stools. These eggs hatch and develop into larvae in the vegetation, where they are ingested as chimpanzees feed

FIGURE 2.2. A chimpanzee named Jilba chews the bitter pith of the medicinal *Vernonia amygdalina* (also known as mjonso or bitter leaf) in Mahale Mountains National Park, Tanzania. Photo by Michael A. Huffman.

on plants. Worm larvae survive much better under wet than dry conditions. Chimpanzees also used medicinal plants more often during the wet season, and the occurrence of self-medication coincided with high worm infections.

Having published a series of papers on his studies, Huffman persuaded not only himself but also the scientific community that chimpanzees can medicate themselves.[25] In his 2012 TEDx Osaka talk, Huffman describes his and Mohamedi's observations of Chausiku as the first scientific observation of an animal actively medicating to cure itself of disease.[26]

WORMS AND VELCRO

In a way, Huffman did not discover anything new. He says so himself when he describes his scientific finding as a *re*discovery of what local people—and chimpanzees—already knew. Animals medicate themselves; and traditional healers all over the world have looked at animals to develop their own cures, likely for thousands of years.

Over a period of twenty years, Mohamedi worked with Huffman as a field guide. But in Huffman's own words, Mohamedi was more than a field guide. He was an equal collaborator in the research and a respected friend. And being descended from a long line of traditional healers, he was an active healer in his local community. In 2000, Huffman learned that Mohamedi's family, and the WaTongwe in general, have long believed that animals medicate themselves, and that people can learn from them.

Huffman remembers the moment well. The chimps he and Mohamedi were following had just ascended a steep mountain, where the researchers could no longer follow. Sitting on a log, they were chatting to pass the time. "I think I was trying to explain to him what it is like on the moon," says Huffman. "That there is no oxygen and that it is very cold. I was trying my best to explain this in Swahili, our language of communication. Then he suddenly mentioned an example of a plant that his grandfather learned to be medicinal from watching a sick porcupine."

As the story goes, Mohamedi's grandfather observed a porcupine with bloody diarrhea dig up and eat the roots of mulengelele (*Aeschynomene* sp.). It was known to be poisonous; people steered clear of it, and they had no medicinal use for it. But the porcupine recovered, and Mohamedi's grandfather decided to make a decoction of the roots. He used it to treat people in his village who were suffering from similar symptoms but were not responding well to his usual treatments. Because his patients were skeptical of the new treatment, he first used it on himself—to demonstrate it would not kill him. The plant is now widely used in western Tanzania, and by some as a substitute antibiotic for the treatment of secondary infections in AIDS patients.

"That was the first time he told me about mulengelele," says Huffman. "I almost fell off the log!" Mohamedi told Huffman about his own observations, and about the stories that were passed to him from his mother, grandfather, and uncles. They credit sick elephants, porcupines, bush pigs, and other animals for their discoveries of treatments for dysentery, stomach upset, sexually transmitted infections, and other ailments.

During the time that Mohamedi worked with Huffman, Mohamedi also developed a new treatment for diarrhea based on observations of chimps who swallow leaves of a plant called mhefu (*Trema orientalis*) to expel the same worms that were treated with *Vernonia amygdalina*. He crushed the leaves, made a concoction, and then treated people with diarrhea. His mother was skeptical because she had never used it as medicine before. But Mohamedi convinced her, and she too began to use it for her patients in the Buhingu area north of Mahale on the coast of Lake Tanganyika.[27]

Even before Huffman was piecing together the evidence for bitter pith chewing, other primatologists had set their eyes on an unusual feeding habit that they thought may be a form of chimpanzee medication. As far back as the 1960s, world-renowned primatologist Jane Goodall noticed that chimpanzee dung sometimes contains whole, unchewed leaves of a plant known as *Aspilia pluriseta*.[28] Scientists Richard Wrangham and Toshisada Nishida described this in more detail in a paper in 1983.[29] Studying chimpanzees in Mahale Mountains National Park as well as in Gombe National Park, also in Tanzania, they observed chimpanzees swallow *Aspilia* leaves whole. A chimpanzee would strip a leaf of the plant, carefully fold it between the tongue and upper palate of the mouth, and then swallow it. Analysis of feces showed that the leaves were passed undigested, and fecal pellets sometimes contained as many as fifty leaves. Wrangham and Nishida noted that *Aspilia* is often used by traditional healers for a range of illnesses, including lower back pain, nerve pain, and high blood pressure. This led to the suggestion that chimpanzees may use *Aspilia* as medicine as well.

Wrangham and Nishida teamed up with chemist Eloy Rodriguez, who reported the antibiotic compound thiarubrine A in the leaves of *Aspilia*.[30] This chemical is highly toxic to pathogenic bacteria, and the researchers hypothesized that chimpanzees swallow the leaves of *Aspilia* as an antiparasitic drug. They also suggested that the careful folding of the leaves and swallowing them whole would protect the leaves from digestive degradation in the stomach, and allow the leaves to deliver the chemicals to the small intestine, where the parasitic worms take residence.[31] However, subsequent chemical analyses were unable to identify thiarubrine A at high enough concentrations.[32] This suggested that leaf swallowing serves a purpose other than delivering toxic chemicals to the small intestine.

Meanwhile, Huffman had collected evidence that leaf swallowing is widespread in chimpanzees, and also in bonobos and gorillas.[33] Moreover, he found that these apes use at least forty different plant species.[34] Comparing the dry and wet seasons,

Huffman also found that chimpanzees swallowed more leaves during the wet season, when infections with that debilitating worm *O. stephanostomum* were more common.[35] Wrangham found similar patterns in Kibale National Park in Uganda, where chimpanzees swallowed leaves a lot more when tapeworm infections were common.[36]

Carefully analyzing all the different plant species that chimpanzees use for leaf swallowing, Huffman noted that they all have one thing in common: their leaves are rough, carrying tiny stiff hairs. Indeed, Huffman and his colleagues would often use leaves of one of these plant species as make-shift sandpaper, to smooth the wooden handles of their machetes. Huffman then realized how chimpanzees benefit from these rough, hairy leaves after a sick spell. Having followed a sick chimpanzee for a day, and having collected her feces, Huffman fell ill with a viral infection. After a three-day recovery period, he finally had a chance to analyze the chimpanzee dung, which he had stored in a plastic bag in the storage room of the field station in Mahale. Astonishingly, he found the bag teeming with wriggling worms that were very much alive. This really put to rest the idea that chimpanzees swallow leaves because they contain worm-killing toxins. Instead, Huffman found some of the worms stuck to the leaves, caught in the hairy surface. Pulling the worms off the leaves, he recalls, made the same sound as undoing a Velcro strip. Later work by Huffman and his colleague Judith Caton showed that in addition to capturing worms, the swallowed leaves cause irritation of the digestive tract, resulting in digestive secretions and quick bowel movements that purge parasitic worms in roughly six to seven hours.[37]

As the years went by, Huffman and others documented bitter pith chewing and leaf swallowing in African chimpanzees, bonobos, and gorillas in more than sixteen sites and twenty-five communities across their entire geographical range.[38] We now know that white-handed gibbons in Thailand swallow leaves as well.[39]

The acts of sucking bitter pith and swallowing leaves by chimpanzees would become the first well-described scientific examples of animal medication. But they were not the last. Chausiku, Huffman, Mohamedi, Wrangham, and others sparked a scientific field of exploration. In the following chapters, we explore the many fascinating ways that animals use medicine to keep themselves healthy. But first, we will ask why animals need medication in the first place.

3

Parasites and Pathogens

In 2016, underwater photographer and diving instructor Kristina Vackova took a stunning photograph of a clown fish gazing at her camera in the Lembeh Strait in Indonesia. Surprisingly, when looking closely at the picture, one can see that there are not two but four eyes staring at you. Indeed, in the fish's mouth resides a marine woodlouse–like creature called *Cymothoa exigua*. This animal bites off the tongue of the fish, attaches itself to the remaining stub, and steals the fish's food when it eats. *C. exigua* is a textbook example of a parasite: an organism that lives inside or on a host, where it uses up resources and damages the host, without providing any benefit at all. As such, it fits right in with the protozoan parasite that infects monarch butterflies, and the intestinal worms that live in the guts of chimpanzees.

The word parasite stems from the ancient Greek *parásitos*, which translates to a person who eats at the table of another. As biologists adopted this term, they must have made the implicit assumption that that person is not an invited companion but an unwanted guest who steals food without giving anything in return. Many scientists like to distinguish parasites from pathogens, which similarly live at the expense of their hosts, and whose name derives from the ancient Greek words *páthos* and *genés*,

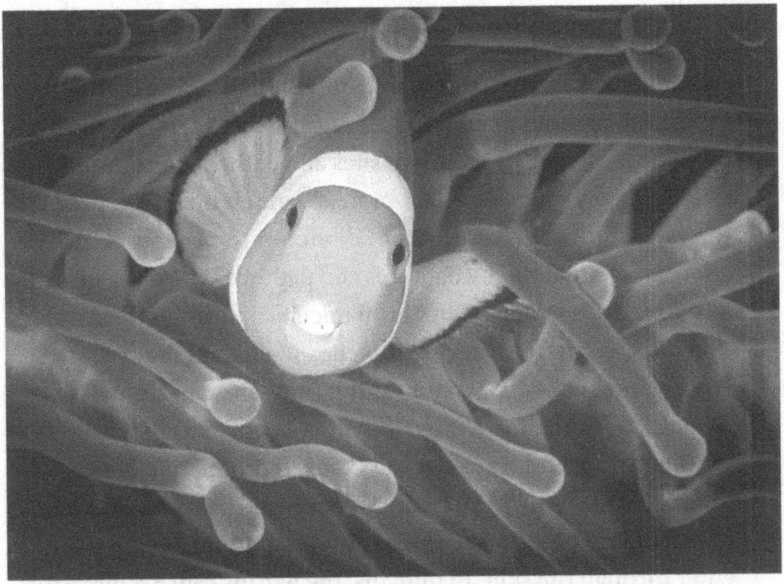

FIGURE 3.1. A clownfish carrying an isopod parasite in its mouth. Lembeh, Indonesia. Photo by Kristina Vackova / Shutterstock.

referring to "suffering, pain," and "producer of." Pathogens are often viewed as smaller infectious organisms, such as viruses and bacteria, with parasites referring to larger ones, such as protozoans, worms, ticks, lice, and fleas. The definitions are not strict, however. Some scientists refer to the one-celled protozoan *Plasmodium*, which causes malaria, as a pathogen, while others call it a parasite. Similarly, some will refer to disease-causing fungi as pathogens, while others call them parasites. I personally do not see the need to distinguish between pathogens and parasites. What matters is that they both cause disease and rely on their hosts for their survival and reproduction. I will use both terms throughout this book and will choose the term that the respective scientists use to describe the organisms they study.

The reason I want to spend some time on parasites and pathogens is that they are exceedingly common and are the prime reason for the existence of animal medication. Some biologists estimate that more than half of all organisms on this earth live a parasitic lifestyle.[1] As such, parasites and pathogens outnumber free-living organisms. What's more, even the mightiest predators, which themselves do not fall victim to other predatory animals, are attacked by parasites and pathogens. Wolves in Yellowstone, for example, suffer from microscopic mites that cause skin disease, and from protozoans that infect their brain.[2]

Left unchecked, parasites and pathogens can do massive damage to their hosts. Schistosome worms sterilize water-residing snails and take over the snails' reproductive organs to produce large numbers of worm offspring. After bursting from the snail, the parasites will search for a mammalian host and enter through the skin. In humans, the worms cause schistosomiasis, a disease characterized by anemia and malnutrition.[3] Nuclear polyhedrosis virus infects caterpillars and turns them into a black goo that contains tens of millions of viral particles. As the goo drips on the plants that the caterpillars feed on, other caterpillars will ingest the viral particles and become infected.[4] House finches infected with a bacterium called *Mycoplasma gallisepticum* experience weight loss and conjunctivitis. This eye infection can be so severe that the birds lose vision and die because of flight accidents. Before they do, though, they scatter bacteria onto bird feeders where other birds pick up the infectious bacteria.[5] It is exactly because parasites and pathogens are so damaging that hosts have evolved a wide range of defenses against infection.

To better understand how parasites and pathogens have given rise to all sorts of defenses in hosts—including medication behaviors—let me give you a crash course on evolution.

OLD BEARDED MEN

As a biology professor, my favorite class to teach is Foundations of Biology—or "intro bio" as my students like to call it. During my first lecture, I show my students two side-by-side pictures of old bearded men. I ask them if they know who they are. Most will recognize the picture of the solemn-looking Charles Darwin, but very few recognize Alfred Russel Wallace, a contemporary of Darwin who came to the same conclusion that life evolved from a common ancestor through a process of natural selection. My next slide shows pictures of two young men. When asking my students if they recognize them, I let the seconds tick by in uncomfortable silence. Some will eventually raise their hand to respond questioningly: "The same guys?" The answer is, of course, a resounding yes. I like to show pictures of a young Darwin and Wallace to clarify to my students that scientists generally make their biggest discoveries when they are young—not much older than most college students.

When Darwin was twenty-two, he set sail on a five-year voyage on the HMS *Beagle* as a gentleman companion to Captain Robert FitzRoy. During this life-changing trip, he visited many different countries and islands around the world, and started to contemplate how species may evolve over time.[6]

When Wallace was in his twenties and thirties, he was in the business of collecting natural specimens, many to sell to zoological gardens. While visiting the Malay Archipelago (now known as Indonesia and Malaysia), he observed unique species on different islands, and likewise came up with the theory that species change and diversify over time.

I talk about Darwin and Wallace to my students to highlight how their theories of evolution and natural selection provide the framework for modern biology. For one assignment, my students compare writings by both scientists from the year 1858. Twenty-two years earlier, at the age of twenty-seven, Darwin had returned from his voyage on the HMS *Beagle*. As the story goes, he visited the Galápagos Islands near the equator and saw different finches on different islands, and—Eureka!—came up with the idea that life evolved from common ancestors and that species diverged from each other through natural selection. This story is overly simplified and incorrect. It was actually mockingbirds and tortoises, not finches, that made Darwin consider evolution; and it took Darwin many years to put all his observations together. Darwin was a meticulous and cautious scientist, and rather than rushing into penning his theory, he decided to spend decades doing research that would allow him to write his ultimate magnum opus on natural selection.

In 1858, Darwin received a letter from Wallace. I like to describe this event to my students as Darwin's "holy shit" moment. Here is a letter from a rather unknown Englishman, who spends his life in the Amazon and Malay rainforests to collect specimens for zoological gardens and collections, while making exquisite naturalist observations. The letter asks Darwin for his opinion on an essay that Wallace wrote on natural selection—during a bout of fever caused by infection with malaria parasites. Darwin realizes that here is his own theory. Is he scooped?

Darwin was friends with many of the leading scientists of that era, so he showed the letter to the geologist Charles Lyell and the botanist Joseph Hooker. Both recommend he provide them with the already-written abstract of his unpublished work, as well as an earlier letter, as evidence that Darwin had independently come up with the same idea. They send these, along with Wallace's essay, to the Linnean Society in London. Spurred on by these events, Darwin rushes to write a lengthy summary of his work in book form, and one year later (in 1859), his *Origin of Species* is published. Darwin actually never finished the ultimate book he had planned. Yet, the *Origin of Species* remains, to this day, Darwin's most famous work and one of the most influential books of all time.

What is most striking about Darwin's and Wallace's writings from 1858 is how similar they are. Both Darwin and Wallace concluded that life was characterized by a struggle for existence, in which individuals compete with each other for food or sexual partners, face harsh climatic conditions, and are attacked by predators and parasites. And in that struggle, only those individuals with particular traits are able to survive and reproduce and are said to have higher fitness.[7] Neither Darwin nor Wallace had a good understanding of genetics, but they clearly understood that those beneficial traits are passed from parents to offspring.

Darwin's and Wallace's conclusions were revolutionary and groundbreaking. As geneticist Theodosius Dobzhansky famously wrote over a hundred years later: "Nothing in biology makes sense except in the light of evolution."[8]

Research over the last 150 years has shown us that lizards inhabiting islands with hurricanes have large toes that enable them to hang on to vegetation.[9] We know that humans living in areas near the equator have evolved dark skin that prevents the breakdown of folate by high levels of ultraviolet radiation.[10] We understand how viruses that jump from other species can quickly adapt to humans and thereby cause pandemics.[11] And we have learned that parasites and pathogens are extremely important for the evolution of species. Because parasites and pathogens are so good at causing disease and death, their hosts experience very strong selection pressure to evolve defenses against these foes, meaning that any trait that results in protection is likely to be selected for and become part of the host's defensive arsenal.

IMMUNE CELLS AND SOCIAL DISTANCING

Hosts have a wide variety of defenses to protect themselves against parasites and pathogens. They include the skin, which forms a barrier for most infectious organisms. They also include the mammalian immune system characterized by T cells, B cells, and antibodies. Most of us are well aware of these cells and molecules, as we tend to stimulate them with lifelong and seasonal vaccines. Insects and other invertebrates possess some rather sophisticated immune responses, too, with antimicrobial molecules, parasite-suffocating pigments, and pathogen-eating blood cells.

Some animals have established long-term relationships with microbes to help them against infections. Aphids harbor bacteria in their bodies that protect them from parasites and pathogens.[12] And some salamanders host bacteria on their skin that inhibit the deadly chytrid fungus.[13] Even plants have a variety of immune responses, including a process called programmed cell death, by which they kill cells infected with pathogens and thereby prevent their spread to the rest of the plant. Interestingly, animals may also kill parts of their own bodies to rid themselves of parasites. In one extreme example, some sea slugs decapitate themselves, severing their own heads from their parasite-infested bodies. Following this parasite purging, the head regrows a whole new body.[14]

Animals also use behaviors to protect against parasites and pathogens. One effective strategy is to avoid becoming infected in the first place. To do so, rainbow trout swim away from eye flukes that cause blindness;[15] great tits avoid nest boxes with lots of blood-sucking fleas; many ants remove sick and dead individuals from their nests; and sheep, horses, and cattle avoid grazing near fresh feces, which are likely to contain infectious bacteria and worms.[16] Birds use their beaks to preen their feathers, removing blood-sucking and feather- and skin-eating lice.[17] Ungulates flick their tails, ears, and heads to prevent biting flies from landing on their bodies.[18] The disgust we and other animals experience when being confronted with feces, body fluids, or stinking water is an emotional response that helps us avoid contact with infectious organisms.[19]

Animals may also avoid contact to prevent disease transmission. As the COVID-19 pandemic took off in early 2020, and no vaccine was available yet, our first line of defense was to avoid contact with others so as to reduce risk of infection. Many of us stayed at home, schools closed, and travel was restricted. Social distancing is not unique to humans. Many animals do it. Some ants infected with fungal pathogens leave their nest hours or days before they die of infection, breaking social contacts with their nest mates and thereby avoid infecting others.[20] In hives infested with parasitic mites, young and old foraging honey bees maintain greater distances between each other than in uninfested hives.[21] Sick vampire bats avoid grooming other individuals, and healthy bullfrog tadpoles, guppies, and finches tend to avoid contact with infected individuals.[22]

So, it is clear that animals have evolved a wide range of behaviors that help them against the threats of parasitism. All these behaviors are extremely exciting to me. But as you can guess, I am especially enamored with animal medication. As I mentioned in chapter 1, we have learned that sick animals can use toxic plants to clear existing infections (therapeutic medication), and that healthy animals can consume medicinal berries to prevent infection (prophylaxis). We have also learned that animals can rub their fur with anti-mosquito repellents (body anointing) and add poisons to their nests to kill blood-sucking mites (fumigation).

At this point it is good to remember that natural selection acts on both hosts and their enemies. Biologists often describe the interaction between hosts and parasites and pathogens as a coevolutionary arms race: because it is in hosts' interest to avoid infection, they will evolve strategies to stop parasites and pathogens from entering their bodies or growing inside them. In response, parasites and pathogens will evolve strategies to overcome host defenses. In turn, hosts will strengthen their defenses, and so on. As a consequence, hosts and their parasites and pathogens are caught in an endless evolutionary cycle where every host defense is countered by a parasite or pathogen strategy to break down or overcome that defense. This endless cycle is often likened to the Red Queen from Lewis Carroll's book *Through the Looking-Glass.*[23] When Alice observes how she seems to stay in the same place while running as fast as she can in Looking-Glass Land, the Red Queen tells her: "Now, here, you see, it takes all the running you can do, to keep in the same place."

As we just saw, animal defenses against infection come in many different forms, including tough skins that prevent parasites from entering the body, and highly evolved immune systems that can prevent infections or stop pathogens from growing inside the body. In response, parasites have evolved sharp mouthparts to pierce tough skin (think of ticks), and many parasites and pathogens have mechanisms by which they avoid recognition by immune cells or even down-regulate the immune system itself. One reason that malaria infections can become chronic is that the parasites ingeniously change their "coats," so they are no longer recognized by the host's antibodies, and the host needs to develop new ones, which takes time.

In addition to piercing skins and avoiding immune detection, many parasites have become masters at altering their hosts' behaviors. They have become puppet masters making their hosts do things that allow the parasites to spread to new hosts and thereby ensure their survival and reproduction. One classic example is the rabies virus. This virus generally enters a host through the bite of an infected animal. It then enters nerve cells to travel up to the brain, where it multiplies in the limbic system, the brain region that controls aggression. The virus also infects the nerves that control swallowing. By reducing the animal's ability to swallow, the virus causes the saliva to build up in the mouth. And as it replicates, the virus fills that excess saliva with countless virus particles. Meanwhile, viral infection of the brain increases the animal's aggressiveness and makes it bite others, thereby ensuring infection of a new host.

Another spectacular example of host manipulation comes from the fungus *Entomophthora muscae*.[24] This pathogen infects houseflies (*Musca domestica*). Infected flies seek out elevated positions and then open their wings as they die to increase the spread of spores as they fall. What's more: the fungus produces a mix of chemicals and alters the chemicals on the female fly's body to create an irresistible—and deadly—bouquet to male flies. Unable to resist their sexual urges, male flies seek out the (dead) infected females and mate with them. The close contact that mating requires provides an excellent opportunity for the fungus to spread from the female cadaver to the unwitting male.

FIGURE 3.2. A female fly infected with a fungal pathogen succumbs to infection; but by manipulating the cadaver into a mating position and mimicking sex pheromones, the fungus tricks a male fly to mate with her. The fungus then infects the male upon contact. Photo by Filippo Castelluci.

As these examples show, parasites and pathogens are experts at host manipulation. Many infected animals behave differently from their uninfected counterparts. This means that when we observe an animal behavior, we cannot simply conclude that the animal is medicating. While evolution has equipped hosts with behaviors to protect themselves, it has also provided parasites and pathogens with sophisticated tools to change their hosts' behaviors for their own benefit. And the trick is to figure out who is pulling the strings. Is it the parasite or pathogen increasing its own success to infect and transmit? Or is it the host animal altering its behavior to protect against infection? Unless we pay attention, we may mistakenly view parasite manipulation as host medication.

So, it takes careful analysis to see who might be winning the evolutionary arms race at any given time. Is it the parasite or pathogen? Or is the host? We will use the next chapter to see how scientists address this question.

4

Beetles and Bulldogs

As we saw in the last chapter, it can be difficult to determine why an animal changes its behavior, especially if it's sick. It takes careful testing to determine whether an animal is actually seeking out medication. I experienced this challenge myself when studying monarch butterflies, and it took careful experiments to conclude that the behavioral changes we observed are indeed a monarch's way to medicate. In this chapter, we'll explore how scientists decide when animals are using medication rather than being manipulated by pathogens or parasites. To begin, we'll explore how I got into this research in the first place.

Many of my students think that scientists have clear, long-term plans to lead them to exciting new discoveries. But this is rarely the case. Many scientific discoveries stem from serendipity and chance, and this is very true for me as well. Sure, as a child, I was interested in nature and animals. My best friend in high school even gave me the nickname "BioBoy"—which was also the name of a compost container that our town provided to its residents. It came as no surprise to anyone that I decided to enroll in a biology program at Wageningen University in the Netherlands. But I had not set my eyes on parasites or butterflies.

Instead, I studied the genetics of a fungus that grows on the dung of rabbits, and measured penis size of dung beetles in the jungle of Malaysian Borneo (I mean, someone had to do it).

I never expected to work on parasites or monarch butterflies, let alone animal medication. My interest in parasites was sparked by a 1998 talk by University of Edinburgh professor Andrew Read during a one-day symposium in Wageningen. It happened to be my twenty-first birthday. Read told a tantalizing story of two different malaria parasites coinhabiting the bloodstream of animals, where the parasites face the same kind of competition for red blood cells—which they eat—as lions do for zebras on the African Savannah. I could not believe it. I had never really thought much about parasites. But here, I heard about the adaptations and struggles of these parasites—how they had to navigate the host's bloodstream and avoid the host's immune system, how they competed with each other for their own survival. That night, I had a chance to talk to Read at dinner. This led to a seven-month internship, after which I decided to pursue a PhD degree in his lab. And so, my twenty-first birthday present was a future career of studying parasites.

CATERPILLARS AND MILKWEEDS

By 2005, I was a postdoctoral scientist at the University of Georgia in the lab of Sonia Altizer, an expert on monarch butterfly parasites. My initial interest in studying monarch parasites was to understand why parasites are—well—parasites. As I mentioned earlier, biologists define parasites as organisms that harm their hosts. But why do they do so? Wouldn't it be more beneficial to harvest resources without harming the host? I spent long days in the lab trying to tackle this question. Monarchs are voracious eaters. It only takes about twelve days for a monarch to develop from egg to caterpillar to pupa. During this time, they multiply their body weight by a factor of three thousand. And to accomplish this, they need food. A lot of food. A single caterpillar, measuring less than a few inches, can devour a milkweed plant that is two feet tall. (I once calculated that to multiply my weight three thousand times, I would need to transform into the mass of two blue whales in a mere twelve days).

To feed my hungry caterpillars, I needed to grow a lot of milkweed. And so I spent a lot of time in the greenhouse, planting seeds, repotting seedlings, fertilizing plants, and trying to kill pests (mostly aphids, thrips, and spider mites). While my lab mates were cheering on the Georgia Bulldogs (the University's football team), I was getting sweaty under the hot Georgia sun (it was okay—I do not really understand the rules of American football anyway). But I liked being in that hot greenhouse, and I was excited to try out some species of milkweed other than the one species—swamp milkweed—I had used until then.

As I was getting more familiar with monarch butterflies and their reliance on milkweed, I was specifically interested in growing tropical milkweed. As I mentioned in chapter 1, tropical milkweed is loaded with toxic chemicals called cardenolides. Swamp milkweed has much lower concentrations of these chemicals. While it was well-known that monarchs use milkweed cardenolides to deter predators, my own experiments showed that they can also reduce infection with the OE (*Ophryocystis elektroscirrha*) parasite. And as I mentioned before, this fact alone made me wonder whether monarchs can actively use toxic milkweeds as medication.

By this time I had teamed up with Mark Hunter, a chemical ecology professor at the University of Michigan. Hunter was instrumental in my research. He not only quantified the cardenolide concentrations in all my milkweeds, he also planted the seed in my brain that cardenolides may act as monarch medicine. In a coffee shop in Ann Arbor, Hunter and I got our notepads out and drew up experiments to test the idea that monarchs can actively use milkweeds as medicine. We had speculated on monarch medication in the discussion of our first article on the medicinal effects of milkweeds,[1] which we had submitted to the *Journal of Animal Ecology*. As is custom in scientific publishing, the editor of the journal sent our manuscript to experts who would comment on our study—a process called peer review. One of the reviewers scoffed at our ideas. There was simply *no way* that monarchs would be able to use plants as medication. The editor made us remove it. We begrudgingly did but never gave up on the idea. A few years later, we would prove that reviewer wrong.

At that time, I had started a position as an assistant professor at Emory University. I had set up my lab and filled the greenhouse with milkweeds. And crucially, a postdoctoral researcher from France joined my team: Thierry Lefèvre had done his PhD research on behavioral manipulation of hosts by parasites. One of the parasites he had studied is a hairworm, which infects crickets. Infecting crickets poses a problem for the worm: crickets are land-dwelling creatures, but the worms mate with each other in water. The solution: the worms manipulate the crickets and make them drown themselves in a body of water.[2] At that point, the worm, which is many times longer than the cricket, wriggles free and exits the cricket carcass in search of a romantic partner. Thus, as Hunter and I were hypothesizing that monarch butterflies could medicate, Lefèvre joined the lab with the hypothesis that the protozoan parasites actually manipulate the monarchs to increase their own fitness.

In the summer of 2009, Lindsay Oliver, an undergraduate student from the University of Pennsylvania, had enrolled in a summer program at Emory, and she joined our team to test whether monarchs can medicate. Our experimental design, first drawn up in that coffee shop in Ann Arbor, was simple. We set up five big walk-in cages in a greenhouse. In each, we placed a medicinal tropical milkweed plant and a nonmedicinal swamp milkweed plant. We would release an individual pregnant female butterfly into the cage and leave her for two hours. We would then count the numbers of eggs she had laid on each species. The result: infected butterflies laid way more eggs on medicinal plants than on nonmedicinal plants.[3] Uninfected butterflies did not.

The important finding here was that infected monarchs made choices that were beneficial for monarchs, not the parasites. Fewer monarch offspring became infected on the tropical milkweed than on the swamp milkweed, and these offspring experienced much lower parasite burdens. Had parasites been manipulating the monarchs for their own benefit, we would have expected the opposite. Our conclusion: milkweeds with high concentrations of cardenolides help monarchs and harm parasites, so the use of those plants by infected monarchs is animal medication, not parasite manipulation. We had discovered that monarchs use drugs to treat their infections.

EVIDENCE AND ANECDOTES

I count myself lucky to study monarch butterflies. Being able to carry out detailed experiments to test specific hypotheses is the dream of many a biologist. And it certainly helped us demonstrate that monarchs can use medication. But for many animals, such experiments are not possible, and we instead rely on observations. So, while experiments may be the gold standard of behavioral research, scientists also use field studies, and to facilitate those, they have assembled a stringent set of conditions to conclude that animals use medication.

The first condition is that the behavior must improve the animal's health or fitness. Demonstrating a benefit for the host is crucial; simply documenting a different behavior could completely fool us, as we can't know if the host is controlling its own behavior or if it's the parasite or pathogen that is changing the behavior of the host for its own benefit. So, scientists must show that the behavior actually helps the host and not the disease-causing parasite or pathogen.

Animals can use medication to reduce the likelihood of infection (as we saw with monarch butterflies) or by killing or clearing parasites (as we saw with leaf-swallowing chimpanzees). In some cases, animals may self-medicate to reduce their symptoms without killing or clearing the infection; this is something we call tolerance. For example, sheep that are infected with barber's pole worms increase their consumption of antioxidants that reduce inflammation, which prevents worm-induced stunted growth.[4] An example of tolerance in humans is the treatment of cholera, which mainly consists of replenishing body fluids;[5] using antibiotics to kill bacteria is only used in severely ill people and is not effective without rehydration.[6]

Another condition is that animals must use external resources: this means that the medicine they use is not made by their own bodies but provided by other species or compounds in the environment. Thus, the use of tough skin or immune cells is not medication, but the use of plants, toxins from other animals, or environmental minerals is. There has been some debate over the years whether medication has to involve toxic substances that would otherwise be harmful to the animal, or whether it can also include food. As I will discuss later, many scientists now agree that animals can use food as medicine,[7] much like a healthy diet is an important part of many forms of traditional medicine in humans.

The animal must also deliberately seek out the medicine. Thus, eating a plant that is not normally part of the diet, or increasing the consumption of a particular food, can be medication. But passively eating a food that happens to contain antimicrobial chemicals is not. The deliberate action can be most easily seen for therapeutic medication, where infection triggers the behavior, and animals start behaving differently from uninfected counterparts. We saw this with chimp Chausiku, who started sucking the bitter pith of a plant when experiencing illness from worm infection. It is harder to demonstrate for prophylaxis and preventative forms of anointing and fumigation, which animals use to prevent infection or bites. Scientists often conclude that animals use prophylaxis when they observe that animals display the behavior more often when infection risk is higher, either because of seasonal changes or because they live in areas with more parasites and pathogens.[8] Capuchin monkeys, for example, use fragrant leaves and toxic millipedes to rub mosquito-repellent chemicals into their fur, and they do so more during the wet season, when mosquitoes are more abundant.[9]

Finally, most scientists studying animal medication agree that the behavior must be costly in the absence of infection. We can think of this in terms of side effects. If you have ever watched a television commercial for a new drug, or read the information sheet with a new prescription, you will know that the negative side effects often seem to outnumber the healing effects for which the drug is prescribed in the first place. In general, we stay away from drugs if we don't need them. This is similar for animal medication. The bitter pith that chimp Chausiku used is toxic at high doses, and best avoided when not needed. Similarly, high doses of the cardenolides in milkweeds that are strongly medicinal against parasite infections in monarch butterflies can also poison monarch caterpillars and shorten butterfly lifespan.[10]

What makes the case so strong for bitter pith use and leaf swallowing in chimpanzees is that scientists studied these behaviors in significant detail. Michael A. Huffman, Mohamedi Seifu Kalunde, and their colleagues spent years carefully documenting that chimpanzees ate medicinal plants that are not normally part of the diet; that they specifically did so when infected with parasitic worms; and that they recovered following the consumption of the medicinal plants. They showed that the plants either contained toxic antiparasitic chemicals or hairy structures that could capture worms and expel them from the gut. And they found that uninfected animals had no interest in the medicinal plants, thus showing that they avoid the negative side effects of the medicine when they don't need it.

But there are many studies that are less clear-cut. I recently received an email from a journalist who was writing a story for *The Scientist* about new research on Indo-Pacific bottlenose dolphins that apparently self-medicate with invertebrates in coral reefs. This research was about to be published in the journal *iScience*.[11] The article told a tantalizing story of dolphins rubbing their skin against corals and sponges (the invertebrate animals, not the artificial scourer sponges we use for doing dishes) in the Egyptian Northern Red Sea. The scientists summarized a very thorough analysis of the antimicrobial chemicals that these invertebrates harbor, and concluded that the dolphins use these chemicals to treat bacterial skin infections. As I told the reporter, this may be medication, but the evidence is not strong enough yet to conclude so. The authors of the study did not actually show that the dolphins benefited from the behavior, that the chemicals cured skin infections, or that the

FIGURE 4.1. A female great bustard examining a male. Female great bustards are picky when it comes to mates, and they prefer to mate with parasite-free males. Photo by Carlos Palacín, Daganzo, Madrid, Spain, April 2004.

dolphins specifically sought out the corals and sponges for medication. Could it be that the dolphins were rubbing the corals and sponges for itch relief—or even for entertainment?

Many studies have shown that lemurs and capuchins bite toxic millipedes and then anoint themselves and each other with the toxic body fluids that the millipedes ooze as their own defense. Lemurs and capuchins most likely use the millipede toxins to deter mosquitoes and botflies.[12] But it remains possible that they use millipedes for other reasons, such as social bonding: the primates often smear the toxins in each other's fur, or rub themselves against each other to obtain some of the toxins.[13] Judging by the writhing masses of monkeys that are drooling and frantically rubbing their own and others' bodies

FIGURE 4.2. A blister beetle oozes toxic body secretions, which may be used by birds to fight infection. Photo by Jaap de Roode.

with pieces of millipedes, the monkeys may well use anointing to cement their social relations.

Or take great bustards, which are large birds that live in open grasslands. Researchers found that males eat more blister beetles than do females during the mating season. These insects produce copious amounts of a chemical called cantharidin. Because the chemical is toxic to worms and bacteria, the researchers hypothesized that the male birds consume the beetles to reduce worm and bacterial infections, and thereby become more attractive partners for their female counterparts, who are very choosy and prefer to mate with males in excellent health.[14] The idea is tantalizing, yet we need more detailed studies to prove that the birds are actually medicating themselves.

So, while there is ample evidence that animals can medicate themselves, there are also many examples of animal behaviors that are rather more anecdotal. Personally, I feel that many *possible* cases of medication are in fact *real* cases of medication. I love the idea that dolphins and bustards medicate. I believe they can, and I think they probably do. It is just that the scientific evidence for those cases has not yet convinced me. But with ongoing studies, it is only a matter of time before we can add many additional animals to the long list of proven animal doctors.

To show how we can move from anecdotes to controlled experiments, let's shift our focus to birds, many of which add toxic plants and substances to their nests. While scientists had long assumed that birds do this to fumigate their nests and kill parasites, it took a series of well-designed studies and a load of cigarette butts to finally prove this idea.

5

Birds and Butts

For many wild species, observational studies are the best or only way that scientists can study medication. It is simply impossible or unethical to experiment on chimpanzees, capuchins, or lemurs in the wild. But some wild species are quite amenable to controlled experiments. As we will see in this chapter, birds have grown into wonderful test subjects when it comes to studying animal medication. The main reason: birds lay their eggs in nests, and often add aromatic plants or other substances to fumigate their nests. Because nests are often easily accessible, researchers can directly manipulate them by adding or removing aromatic substances to study the effect on disease. And—they can even add parasites to determine how the birds respond.

Humans use fumigation to kill pests, parasites, or pathogens in buildings, operating rooms, or food supplies such as grain and produce. For example, pest exterminators may fill an entire building with noxious gas for several days to kill termites. And a common method to control disease-transmitting mosquitoes is to spray houses or gardens with clouds of insecticides.

Animals also add all sorts of materials to their homes. For example, fruit bats (*Cynopterus sphinx*) in India incorporate aromatic plants into their tent roosts,[1] dusky-footed wood rats (*Neotoma fuscipes*) add California bay leaves (*Umbellularia californica*) to their sleeping nests,[2] and chimpanzees incorporate branches of Uganda ironwood (*Cynometra alexandri*) into their night beds.[3] Because many of these added materials are toxic, scientists have often speculated that animals use them to kill parasites and pathogens. Thus, when we see blue tits (*Cyanistes caeruleus*), russet sparrows (*Passer cinnamomeus*), and Bonelli's eagle (*Hieraaetus fasciatus*) add antibiotic and insecticidal plants such as wormwood and pine needles to their nests,[4] it is tempting to conclude that they do so to kill bacteria, mites, and botflies.

The study of animal fumigation took off at about the same time that Michael A. Huffman and colleagues were first observing therapeutic medication in chimpanzees. In 1985, Larry Clark and J. Russell Mason wrote a paper to demonstrate that many birds line their nests with plants to ward off parasites and pathogens.[5] At the time, scientists had competing ideas for why birds incorporate green foliage into their nests: maybe the green leaves help with thermoregulation; or they camouflage the nests in trees and bushes; or maybe male birds decorate their nests with plants to attract females.

In a field study of European starlings (*Sturnus vulgaris*) in Avondale, Pennsylvania, Clark and Mason found that male starlings are very particular about the plants they collect: they had a strong preference for wild carrot (*Daucus carota*) and fleabane (*Erigeron philadelphicus*). Because Clark and Mason realized that these plants are commonly used by humans as herbal medicine, they hypothesized that starlings use them to protect their chicks against parasites and pathogens. To test that idea, the researchers manipulated starling nests by adding wild carrot leaves.[6] This reduced the number of blood-sucking mites and resulted in healthier chicks that suffered less from anemia.

The power of Clark and Mason's study lies in its experimental design: rather than observing a correlation between the presence of plants in nests and the health of chicks in those nests—as most other studies on fumigation have done—they specifically manipulated nests, adding plants to show that they protect against mites and increase chick health.

While this was a major step in the study of fumigation, there remained one problem, though: even if adding plants changes the number of parasites and the health of chicks, is that why the birds add the plants in the first place? Or do the birds add the plants for another reason, and it is just fortunate that the plants happen to kill mites and improve chick health at the same time? To get at that question, scientists would need to do experiments in which they manipulated not only the presence of plants but also the presence of parasites. This would then allow them to ask whether birds can respond to the presence of parasites by specifically adding medicinal plants.

Luckily, scientists have now done studies exactly like that, conclusively demonstrating that birds use fumigation to kill parasites. Ironically, it took a lot of cigarettes to provide that evidence. To explain how, let me take you on a trip to Mexico City.

URBAN MEDICATION

I have visited the capital of Mexico several times, always on my way to visit the monarch butterfly overwintering sites. But on a cloudy day in the summer of 2022, I am still struck by the sheer size of the city. Having grown up in the Netherlands, with a total population of around 17 million people, the city of 22 million people feels overwhelming to me. With this many people, the traffic can be bad. And indeed, my Emory University colleague

Miguel Reyes, who has graciously agreed to join me on this trip to his home city, and I hit some major traffic jams.

We are quite delayed when we finally make it to the Tienda (shop) at the campus of UNAM, the Universidad Nacional Autónoma de México. Like the city, the university campus is massive, but it provides a nice break from the bustling city, with its many trees and green natural areas that connect the different schools and institutes. The entrance to the Tienda is a convenient meeting place, and we look out for Monserrat Suárez Rodríguez, a professor at the Facultad de Estudios Superiores Iztacala (Faculty of Higher Studies Iztacala) who discovered that house finches (*Haemorhous mexicanus*) and house sparrows (*Passer domesticus*) use cigarette butts as medication.

Having met Suárez Rodríguez on video calls before, I quickly recognize her. As we meet up, she introduces us to master's student Helena Maldonado Hernández, who studies the effects of noise pollution on the breeding success of vermilion flycatchers. Together, we walk over to Suárez Rodríguez's canary-yellow Toyota Prius, so she can drive us to some of her field sites. At the car, Suárez Rodríguez and Maldonado Hernández try to remove the toddler car seat in the back, but give up after a few minutes. (As I watch, I am reminded of the many times that I used manuals and YouTube videos to install or remove the car seats of my own toddlers some ten years ago). So, while Suárez Rodríguez and Maldonado Hernández take the front seats, Reyes and I share half of the back seat and are awkwardly contorted with some limbs sticking out of the open window. Luckily, the ride is a short one to the Instituto de Investigaciones Filológicas (Institute of Philological Research; philology is the study of languages), and we get out to walk round the building to the library in the back. On our way, Suárez Rodríguez points at the littered cigarette butts that line the sidewalk.

The library sits at the back of a large, grassy area, adorned with shrubs and trees with bright red berries. The building itself looks like a wall of glass, some three stories high, and held together by white-painted metal pillars and cross beams. Where the pillars meet the cross beams, there are small openings that give access to the insides of the metal beams. And there, in those hollows, house finches and sparrows build their nests.

We can see twigs and pine needles dangling from the openings in the metal. Parent birds fly to and fro, and there is a lot of singing and chirping, from both parents and chicks. There are at least thirty nests on the building, all built inside the well-protected, hollow metal beams. Suárez Rodríguez explains that the sparrows mostly build nests in the upper metal beams, while the finches make their nests at the lower level. They seem to coexist happily. Using binoculars, I note a male finch sitting on the metal beam above a nest. He has an astonishingly bright red head. The male finches I know from the Atlanta area tend to be more drab.

"Those male finches just care about singing and looking beautiful," says Suárez Rodríguez. "They do not help much with raising the chicks."

Male sparrows, on the other hand, are helpful fathers, putting in their share of nest-building and childcare.

When it comes to that nest-building, the sparrows and finches have found an ingenious way to protect their chicks from the parasitic mites, lice, and ticks that tend to take up residence in the birds' nests. And that protection is as much an adaptation to urban life as the building of their nests on a university building. While the birds use many natural materials, such as twigs, grasses, pine needles, and spider webs, they also happily incorporate human-made substances—including cigarette butts.

Suárez Rodríguez points at a door at the back of the library. "Researchers who work in the library come through that door to take breaks here. They smoke and litter their cigarette butts on the ground."

At the moment, there are not as many cigarette butts as there used to be, and we see fewer than we saw on the sidewalk leading up to the library. But that, says Suárez Rodríguez, is because UNAM went on a COVID-19 lockdown for two years. With researchers doing most of their work from home, there were fewer people here to discard their cigarette butts. Before the pandemic, however, there was a vast supply of cigarette butts strewn across the lawn. Finches and sparrows do not venture far to build their nests, and the piles of cigarette butts provided an easy building material for the birds.

Suárez Rodríguez first studied the nests when she was an undergraduate student in Constantino Macías Garcia's laboratory. Macías Garcia is an ecologist at UNAM who studies the effects of pollution, noise, and urbanization on animal behaviors. His previous research had shown that birds in cities have different songs than those in more natural areas—so they can be heard over the noise caused by traffic and other human activities. Together, Suárez Rodríguez and Macías Garcia decided to study how birds incorporate human-made materials into their nests and what the consequences are for their health.

"We had this idea that they use synthetic strands or plastics, and that the chicks may get tangled in them," says Suárez Rodríguez.

When examining the nests to see if they contained plastic, Suárez Rodríguez soon discovered that many of the finch and sparrow nests contained large amounts of white fibers. At first it was not clear what they were, but when rain turned the nests wet, Suárez Rodríguez smelled smoke.

"I was very curious when she told me these are cigarette butts," remembers Macías Garcia.[7] "'Where?' I asked. 'I only see little fibers.' But then she made me smell them."

So, it is not like the birds simply pick up cigarette butts and dump them in their nests. In a very elaborate process, they collect the butts, remove the outer husk and then weave the individual fibers into the lining of their nests.

"They take the time to open the butts and remove the fibers," says Suárez Rodríguez. "That told me that the butts are useful for them."

To find out just how useful, Suárez Rodríguez monitored 28 house sparrow nests and 29 house finch nests.[8] After chicks fledged, she collected each nest in a sealed plastic bag and took them to the lab. There she placed each nest in a Berlese-Tullgren funnel: a contraption where a bright warm light is placed above the nest and a funnel with a vial of ethanol is placed underneath. The light and heat drive mites out of the nest toward the funnel and into the ethanol so they can be easily counted. She also dissected each nest to count the number of cigarette butts. Nests contained an average of 8 to 10 butts, but some contained as many as 48. And: nests that contained more cigarette butts harbored fewer mites, suggesting that the cigarette butts repel the mites.

To test this directly, Suárez Rodríguez bought a 400-pack of filter cigarettes and smoked half of them with an artificial smoker (basically a tube connected to a flask that creates a vacuum to pull air through the cigarette). She then placed smoked and unsmoked filters—combined with a warm resistor that attracts mites—at opposite sides of bird nests and found many more mites moving toward the unsmoked butts. This showed directly that mites avoid smoked butts. This aversion to cigarette butts reduces the number of parasites in the birds' nests and helps the birds. Suárez Rodríguez followed up her bird studies as a graduate student and found that greater numbers of cigarette butts not only correlate with fewer parasites but also increased bird survival.

THANK YOU FOR SMOKING

Suárez Rodríguez's studies showed that smoked cigarette butts can repel parasites and that this is beneficial for the birds. But that does not necessarily mean that the birds collect the butts specifically to get rid of parasites. When Suárez Rodríguez and Macías Garcia first found the cigarette butts they hypothesized that the birds use them as a nice fluffy material that keeps the nest warm. (Indeed, as I am inspecting the nests with my binoculars, I see a sparrow fly to its nest with a big piece of some white fluffy material—I can't make out what it is, but it definitely looks like human trash). To test whether birds collect cigarette butts specifically to reduce parasitism, she needed to do a controlled experiment. And to understand that experiment better, we leave the library of the Instituto de Investigaciones Filológicas, and squeeze back into the car to drive to the Instituto de Ecologia.

After a short drive, we get out and make for a lawn with several picnic tables. We are lucky. A couple of people have just finished their lunch, allowing us to take over their table and transform it into a scientific lab. As we sit down, Suárez Rodríguez gives us green examination gloves, then opens a sealed plastic bag to show us a two-year-old sparrow nest. The nest is a beautiful and complex structure of twigs, pieces of grass, and pine needles. We also note some pieces of paper and blue string. And then there are the cigarette butts. One is kind of obvious, although the husk is gone, and the only visible part is

FIGURE 5.1. Monserrat Suárez Rodríguez shows a house sparrow nest, in which the parent birds have incorporated cigarette butts to keep blood-sucking parasites away from their chicks. Photo by Jaap de Roode.

the remaining white fibers. Then she pokes in the middle of the nest to show a whole dense layer of the same fibers. I can see now that it must have taken some detective work to figure out that the fibers are indeed cigarette butts.

Suárez Rodríguez goes on to remove the top layer of the nest, which includes the butts. This is the lining of the nest. Its main role is to keep the eggs and chicks nice and warm. Being able to replace the lining with an artificial one was essential for her next experiment. It took Suárez Rodríguez some time, but after several years, she figured out how to remove the lining and replace it with an artificial lining that she made herself.

"The birds easily reject the nests when they are disturbed," she says.

She learned that if she manipulated the nests *after* egg laying rather than before, the birds wouldn't reject them. She also learned that she needed to use materials that the birds naturally collect: branches, sticks, and pieces of grass.

After examining the nest, we walk over to the institute's parking lot, which is located under a ramada-like structure that is covered with solar panels. It is very clever: not only does the structure provide shade for the cars, reducing the need for air conditioning, but also the solar panels capture renewable energy. Underneath the panels are light fixtures, and as we are looking at them, it is clear that these provide another great place for house sparrows and finches to build their nests. Maldonado Hernández has brought a little mirror attached to a long pole that she can position over the nests. As she holds it over a finch nest, we can see two beautiful light green eggs. Because the nests are easy to access here, this parking lot provided a great place for Suárez Rodríguez's next experiment. She located 32 house finch nests with eggs, then carefully removed the eggs and placed them in a bowl with cotton. She then removed the natural lining and replaced it with an artificial lining—without any cigarette butts. She either left the lining as is, added live ticks, or added dead ticks. After that, she placed the eggs back in the nest, making sure to put the nests in the exact locations where she had found them.

There was a good reason for choosing ticks for her experiment.

"There are a lot of things living in the nest," says Suárez Rodríguez. "Take mites: some are parasitic, but others could just be living there in the nest. But for ticks, we know they have to puncture the bird skin to suck blood. So we are sure they are parasites, and that they are detrimental."

At the end of the breeding season, as the chicks had fledged, she dissected each nest and counted the number of cigarette butts. And amazingly, she found that the nests with live ticks contained a lot more cigarette butts than those with dead ticks or no ticks.[9] This showed that the house finches actively collect more butts when they are experiencing live, blood-sucking parasites.

Whether the birds specifically seek out smoked—as opposed to unsmoked—cigarette butts and how exactly the birds know that the cigarette butts are helpful to them remains unknown. Suárez Rodríguez and Macías Garcia placed smoked and unsmoked cigarette butts in boxes to see which the birds preferred. Although the birds collected the smoked ones faster, they ended up collecting both. Based on these results, Suárez Rodríguez thinks that the cigarette butts may have been so abundant that the birds simply picked them up. And perhaps they started associating the use of cigarette butts with the lower abundance of parasites in their nests. Alternatively, it is possible that one bird discovered the beneficial effects of the cigarette butts, and other birds copied the behavior.

"They are very social, and definitely learn from each other," says Suárez Rodríguez.

As we know, smoking cigarettes is bad for human health. Likewise, cigarettes are not harmless for birds. In yet another study, Suárez Rodríguez and Macías Garcia found that chicks raised in nests with more cigarette butts carried some abnormalities in their blood cells, which may reduce their survival later in life.[10] The antiparasitic effects of cigarette butts are most likely due to nicotine, a potent toxin that paralyzes and kills many animals. Tobacco plants are well-known for their antiparasite utility: organic rabbit and poultry farmers in British Columbia add tobacco plants to nest boxes to treat diseases.[11] But nicotine is not the only chemical in smoked cigarette butts: they also include heavy metals and phthalates (chemicals that are added to plastics to make them more flexible, transparent, or durable). And these are likely responsible for the cell abnormalities that Suárez Rodríguez and Macías Garcia observed.

The filters that constitute cigarette butts are made of plastic fibers, and these can take fourteen years to degrade in the environment. Cigarette butts form a serious environmental problem. In 2019, over 1 billion smokers smoked some 6 trillion cigarettes, 4.5 trillion of which were discarded as litter.[12] These butts get carried on surface runoff and enter waterways and oceans. As the butts make it through the environment, their plastic and toxins leak into the earth and water. Studies have shown that this pollution kills brine shrimp, impairs movement of snails, makes it harder for bivalves to dig burrows, lowers reproduction of copepods, reduces germination of white clover, and makes mice less able to escape from predatory snakes.[13] So, despite the health benefits experienced by house finches and sparrows, cigarette butts are bad news for the environment. And of course, they are bad for human health, which is why public health campaigns have focused on trying to get people to quit smoking altogether.

It seems counterintuitive that the house finches and sparrows on the UNAM campus obtain their medicine from a behavior that is so detrimental to human health and nature. And that makes me wonder: If people at UNAM stopped smoking altogether, would this be bad news for house finches and sparrows? What would happen to their supply of medicine?

"I think they are just going to use something else," says Suárez Rodríguez.

Macías Garcia agrees. "A key element of the success of birds living in cities is their adaptability. If they cannot find cigarettes, they can use other materials. They could use *Nicotiana* plants. Many people have aromatic plants on their window sills that they use for cooking, and the birds can use those too."

For the time being, however, the birds seem to do just fine with the available cigarette butts. It is ironic that humans have helped finches and sparrows by providing them with a novel form of medication. I am saying ironic, because the usual way in which humans have affected animal medication is by taking away the naturally occurring medicines that animals like to use. Humans have been experts at domesticating animals, including sheep, goats, bees, and cats. As I will discuss later, this domestication has taken away some very important ways that animals medicate themselves, with severe consequences for animal welfare, drug resistance, and economic sustainability.

Luckily, as we are learning more about animal medication, we are also finding new ways to reintroduce medication in agricultural animals and pets. And as we will see, movements are underway to provide livestock with diets that include medicinal plants; beekeepers are developing new hive boxes that bees can fill with antimicrobial resin; and veterinarians are providing advice on diet and lifestyle changes that allow our pets to optimize their own health.

For now, I want to stick with the theme of experiments, though. When Suárez Rodríguez and Macías Garcia created artificial nests, seeded them with ticks, and then saw birds collect more cigarette butts, they not only showed convincing evidence for fumigation by finches and sparrows but also illustrated the power of controlled experiments. Although such studies are impractical or unethical for many species, they have turned out to be great tools to demonstrate medication in six-legged creatures, animals that few people would have predicted to be effective doctors. As we will see in the next chapter, scientists are increasingly turning to insects to study animal medication.

6

Ants and Aliens

Apes and monkeys tend to steal the show when it comes to animal medication. Scientists have found that no fewer than twenty-five different primate species across twenty-six different countries around the world use plants with well-described medicinal properties in the wild.[1] Primates also use millipedes and clay as medication, and recent reports suggest that they even rub insects into wounds, possibly to reduce skin infections.[2]

But as we have seen, medication isn't limited to primates. Birds and monarch butterflies also medicate. And it turns out monarchs aren't the only insects that do so. While many people think of insects as simple organisms, they are in fact complex and highly evolved, and we shouldn't be surprised that they have evolved medication behaviors. Insects are the most diverse group of animals on earth, with an estimated 5 million species.[3] Insects have also been around for much longer (some 479 million years[4]) than mammals and birds, which have been around for less than 200 million years.[5] Given their diversity and age, it would be utterly surprising if they had not evolved some intriguing medication behaviors.

Insects are easy to maintain in the laboratory in large numbers, so they are well-suited for controlled experimental studies.

And because of that ease of experimentation, insects are becoming more and more important as test cases of animal medication. To see how, let's start by going to the Jura Mountains in Switzerland.

The landscape in the Jura Mountains is dominated by beautiful rolling hills covered in spruce forests. Among those trees one can find huge mounds, many exceeding the height of a small child. In the episode "Supersocieties" of the 2005 documentary series *Life in the Undergrowth*, Sir David Attenborough can be seen beside a mound that is almost as tall as him and many times wider. The mounds are the nests of millions of wood ants (*Formica paralugubris*).

As wood ants build their nests, they mix pieces of resin with spruce needles, bits of wood, and soil to create the walls of their chambers and passageways. Resins are sticky secretions that some plants produce to protect themselves against pathogens and insects and that help them recover from wounds. The resin from conifer trees consists of different types of terpenes that are toxic and repellent to microbes and to insects such as bark beetles (*Dendroctonus*) and pine shoot moths (*Rhyacionia buoliana*). One famous example of resin is the "tears of Chios," the droplets of resin that mastic trees on the Greek island of Chios produce; this resin is often used as gum and has been used as human medicine for thousands of years to treat a whole series of afflictions, including inflammation and stomach ulcers.[6]

While the average hiker may not pay much heed to the pieces of resin in the ant nests, the resin has been known by ant experts and locals for a long time. In *The Natural History of Ants*, written sometime around 1742, French entomologist and writer René Antoine Ferchault de Réaumur mentions that people would regularly collect the resin from ant nests and burn it as incense in their homes. But what do the ants use the

FIGURE 6.1. Wood ants collecting medicinal tree resin. Photo by Simon Williams / Minden Pictures.

resin for themselves? This was a puzzle for Philippe Christe, a researcher at the University of Lausanne who grew up in the Jura Mountains. Ants don't eat resin, and they don't need it as a building material. Yet they put in a lot of work to fill their nests with the material. The ants collect pieces of dried-up resin from spruce trees. Nests can contain up to twenty kilograms of the material, and the collected resin can account for more than 40 percent of the total nest materials.

At the time Christe became puzzled by the ants, his primary area of study was starlings. As we saw earlier, these birds often incorporate fresh plants into their nests to repel ectoparasites.

If birds fumigate their nests with plants, could ants be doing the same with resin? One of Christe's colleagues and friends, also a researcher at the University of Lausanne, was Michel Chapuisat. Chapuisat studied wood ants, and one day they discussed their resin-collecting behavior. Putting their minds together, they quickly came to the conclusion that the ants may be using the resin as medicine.

To test whether ants medicate with resin, Christe and Chapuisat carried out a series of experiments. In the first,[7] they created twenty experimental ant colonies in the lab. They added resin to ten of them and left the other ten resin-free. They then quantified microorganisms, and found that resin-containing nests had fewer bacteria and fungi. They also found that resin inhibited the growth of *Pseudomonas luteola*, a common pathogenic bacterium in ant nests. In a second experiment,[8] the researchers studied two lethal ant pathogens: a bacterium called *Pseudomonas fluorescens* and a fungus with the name *Metarhizium brunneum* (formerly named *M. anisopliae*). They placed adult ants in petri dishes with the bacterium or the fungus, and added resin to some of them; they did not add resin to other dishes. They found that the resin protected the ants, increasing their survival by some 20–40 percent. In a similar way, they demonstrated that resin protects ant larvae from lethal fungal infections.

Having shown that resin reduces the abundance of pathogens in ant nests and protects ants against bacterial and fungal disease, Christe and Chapuisat then wanted to know if wood ants deliberately collect resin or if the antimicrobial effect is simply coincidental. So they went to the mountains, located ant trails, and placed trays with pieces of resin, twigs, and stones—a buffet from which the ants could choose. They repeated the experiment during different times of the year and found that the ants preferred the resin over the other items in spring and summer, but not autumn. Spring and summer are the times when ants are primarily reproducing, so Christe and Chapuisat concluded that the ants were specifically collecting medicinal resin during the time of year when they produce brood (babies).

They found similar results for experiments in which they replaced the pieces of resin with pieces of wood impregnated with turpentine (a chemical extracted from resin), implying that it was the chemical compound the ants were collecting, not just the resin itself.[9]

"They use the resin in a very precise way," says Christe.[10] "They place the resin near the brood in experimental nests."

A follow-up study showed that when baby ants are present in a nest, ants collect a lot more resin than when babies are absent.[11]

Finally, the Swiss researchers did a lab experiment in which they infected some ant colonies with that lethal fungus *Metarhizium brunneum* while leaving others uninfected. They then provided the ants with choices between resin and stones, predicting that the infected ants would have stronger preferences for resin than the uninfected ants. Surprisingly, however, they found that infected and uninfected ants equally liked resin. This suggests that the ants use resin to prevent future infections, rather than treat existing infections. By collecting resin, wood ants keep their brood chambers clean and free of disease.

"It is a remarkable adaptation," Chapuisat says.

SIX-LEGGED HEROES

Wood ants are not the only remarkable insects. Researchers have found that, when surrounded by parasitic wasps, *Drosophila suzukii* fruit flies prefer to lay eggs on diets that contain higher concentrations of atropine.[12] Atropine is a toxic alkaloid that is common in deadly nightshade plants (*Atropa belladonna*), and when the offspring larvae of the fruit fly mothers consume the chemical, it kills parasitoid wasp larvae. Other biologists have found that honey contains antimicrobial chemicals[13] and that honey bees can choose between different types of honey to alleviate parasite infection.[14]

At the 2022 International Congress of Entomology in Helsinki, Finland, researchers devoted a whole symposium to medication in insects. Here, scientists presented evidence that infected ants (*Formica fusca*) medicate themselves by eating food with high concentrations of reactive oxygen molecules that are harmful to fungal parasites.[15] Others showed that bumblebees can use chemical compounds in the nectar of linden trees (*Tilia spp.*) and strawberry trees (*Arbutus unedo*) to reduce gut infection with trypanosome parasites.[16] Intriguingly, the bees either change the chemicals themselves, or let the microbes in their guts alter the chemicals, to make them more potent against the parasites. Similarly, a 2017 study showed that wood ants can make the antifungal tree resin they collect more potent by adding formic acid, which they produce in their venom glands to scare off predators and nest intruders.[17]

It is no accident that insects have risen to prominence when it comes to research on animal medication. Insects tend to be small and easy to handle in labs. They reproduce easily in captivity and can be kept at large numbers. And—research on insects is not subject to the same restrictions as research on vertebrates, including primates. Beyond adding more and more evidence that animals can—and do—medicate, studies on insects are also helping us understand just how chemicals from plants and other organisms help animals maintain their health.[18]

In many cases, the chemicals seem to directly interfere with the parasites and pathogens.[19] Chlorogenic acid from tomato plants can bind to viral particles, which inhibits their infectivity to corn earworm caterpillars (*Helicoverpa zea*).[20] And chemicals from noni fruit can kill parasitoid wasps of fruit flies from the Seychelles islands.[21] In other cases, plant chemicals seem to boost the insect's immune system. In tobacco budworm caterpillars (*Heliothis virescens*), the consumption of cotton leaves rich in chemicals called peroxidases increases the shedding of virus-infected gut cells.[22] And in tobacco hornworm caterpillars (*Manduca sexta*) infected with bacteria, chlorogenic acid from tobacco plants (*Nicotiana attenuata*) increases the numbers of immune cells and caterpillar survival.[23]

ALIEN FLIES

Insects have not just expanded the A–Z list of medicating animals. They have also helped us better define what animal medication really is. As I mentioned earlier, scientists now commonly accept that one of the conditions for animal medication behaviors is that medicines are costly or harmful in the absence of infection. To see how insects helped solidify this idea, let's look at woolly bear caterpillars.

In 1993, Michael Singer, now a professor at Wesleyan University in Connecticut, was puzzled by the behavior of fuzzy caterpillars called woolly bears. At the time, Singer was a graduate student at the University of Arizona. He had joined the lab of entomologist Elizabeth Bernays, an expert on insect–plant interactions. Much of the lab's research focused on the feeding behavior of grasshoppers, but Singer's first love was butterflies and moths. So, while his lab mates were following grasshoppers, Singer was chasing woolly bears.

When tracking through his field sites, he noticed that the caterpillars readily move from plant to plant, eating cheeseweed

FIGURE 6.2. A woolly bear caterpillar on a common fiddleneck plant (*Amsinckia intermedia*), which contains antiparasitoid pyrrolizidine alkaloids. Photo by Michael S. Singer.

(*Malva parviflora*), then threadleaf groundsel (*Senecio longilobus*), then ragweed (*Ambrosia confertiflora*), and plantain (*Plantago insularis*). They move so much that they may easily eat leaves from five plant species per day. For those not familiar with the eating habits of caterpillars: this is quite unusual. Monarch caterpillars, which I study myself, are specialized eaters of milkweed and hardly move between plants at all.

"I thought they were doing it to get a good nutritional diet," says Singer.[24]

But when he did experiments in the lab, he found that the caterpillars did not grow any better on mixed diets than on diets consisting of a single plant species. Thus, there had to be another reason for the caterpillars switching their diets all the time.

As Singer was bringing in caterpillars from the field into the lab, he found that not all of them survived long enough to turn into adult moths. Instead, he found something that would give most people nightmares: many of the caterpillars succumbed as fly maggots burst out of their bodies. The flies in question are known to biologists as tachinid flies. And while they may look like normal house flies, they are anything but normal. Instead of buzzing around windows and laying their eggs in scraps of meat, tachinid flies lay their eggs on caterpillars. Baby maggots hatch from these eggs and eat their way *into* the caterpillar body. There, they gorge on the body fluids and tissues of the caterpillar until they are ready to pupate. At this point they eat their way out of the caterpillar, leaving it dead in their wake. Tachinid flies are a bit like the creature in the classic 1979 movie *Alien* that burst from the chest of spaceship Executive Officer Kane, killing him in the process.

Singer collected caterpillars from ten different field sites in Arizona. In some sites he found that one in ten caterpillars harbored the flies. In others, a staggering 90 percent of the caterpillars were parasitized. And that made him wonder: Are the woolly bears switching their diets to make sure they consume antiparasitic medicine in addition to food?

The antiparasitics that Singer focused on are called pyrrolizidine alkaloids. In humans and livestock, these chemicals cause acute liver toxicity, abdominal pain, nausea, vomiting, and diarrhea. (An easy way to find plants with these chemicals is to visit a well-grazed pasture and collect the plants that have been avoided by grazing animals). Together with colleagues, Singer realized that many of the plants that woolly bears eat have these chemicals. What's more, woolly bears often concentrate these alkaloids and store them in their bodies,[25] which makes the caterpillars toxic to predators.

FIGURE 6.3. A monarch caterpillar has been consumed from the inside out by three tachinid fly maggots. The mother fly laid her eggs on the caterpillar; baby maggots hatched from these eggs and ate their way into the caterpillar body, after which they consumed the caterpillar's insides before eating their way out. Photo by Jaap de Roode.

It was a natural question to ask whether the alkaloids can act as medicine too.[26] So, as Singer moved to Wesleyan University, he and his colleagues carried out a series of experiments to test this idea.

"I was skeptical at first," he says. "It just seemed quite implausible."

Like so many other scientists, Singer did not quite believe that a small insect, like a caterpillar, could have behaviors that we normally ascribe to animals that have big brains and are much like us. But as he thought about it, the idea started to make sense. He already knew that insects are masters at assembling their diets to ensure they get all the nutrients they need. Was it really that much of a stretch to think that they could also make medicinal choices?

Singer was one of the first scientists to suggest that for a behavior or substance to be considered medication, it must be costly to individuals that are not sick.[27] That made sense from his training in evolutionary biology.

"If there are no costs," he says, "the animals would simply use the medicine all the time, but we don't see that."

The idea of costs also resonates with the way humans use medicine. As I mentioned earlier, the number of side effects and risks of drugs usually outweighs the benefits, and we do well to avoid drugs when we do not need them.

Together with a student, Singer created two artificial diets for the caterpillars. One contained pyrrolizidine alkaloids, the other did not. They fed these diets to caterpillars infected with fly maggots and found that the alkaloids increased caterpillar survival: while only about 40 percent of infected caterpillars survived on the alkaloid-free diet, the diet with alkaloids increased survival to 60 percent. For uninfected caterpillars, however, the alkaloid diet reduced survival from 100 percent to 80 percent. Thus, as he expected, alkaloids act as medicine, but they also carry heavy costs.

To demonstrate that the caterpillars can actively use alkaloids to cure themselves, Singer offered both diets simultaneously to four different groups of caterpillars: caterpillars that were uninfected, and caterpillars that were infected with one, two, or three maggots. He then monitored how much of each diet the caterpillars ate throughout their life. Interestingly, when caterpillars suffered infection with two parasitoid maggots, they had a strong preference for the alkaloid food. Uninfected caterpillars did not have such a preference, and neither did caterpillars with one or three maggots. This suggests that for singly infected caterpillars, the costs of alkaloids outweigh the benefits (and caterpillars can use their immune system to battle the single maggot instead); for those infected with three maggots, the caterpillars were completely overwhelmed, and there was simply no point to medicate any longer. But when infected with two maggots, the caterpillars could avoid death by supplementing their own immune system with the deadly alkaloids. A fine example of therapeutic medication.

As I mentioned, Singer's experiments really cemented the idea that animal medication involves the use of substances that can protect against disease but can also have harmful side effects. Another way to say this is that many forms of animal medication involve the use of toxins. When an animal is infected, these chemicals provide them with protection because the benefits from harming the parasites and pathogens that infect them outweigh the harm done to the animal itself. When uninfected, the toxins do more harm than good, and are best avoided. This is as much true for the pyrrolizidine alkaloids used by woolly bear caterpillars as it is for the steroid glucosides used by chimpanzees, the cardenolides used by monarch butterflies, and the nicotine and other cigarette chemicals used by birds.

The fact that so many animals use toxins to medicate themselves raises an important question: why are toxins so readily available in nature? This question is the topic of the next chapter.

7

Poisons and Proteins

On a late-summer day in 1991, two mountaineers from Nuremberg, Germany, were on their way back from a hike in the Ötztal Alps, on the border of Austria and Italy. As they made their way down a mountain ridge, they stumbled upon the remains of a frozen human being. Initially misinterpreting the find as a recently perished mountaineer, scientific research quickly revealed that the remains actually belonged to a man who was some 5,300 years old. Ötzi the Iceman, as a journalist named him, had been mortally wounded by an unknown human adversary.[1] Upon his death, dry, windy, and cold conditions mummified him. A glacier then preserved him for more than five thousand years.

What was remarkable about the find is that Ötzi provided a frozen snapshot of daily life in the Neolithic. Until Ötzi's discovery, most archaeological finds taught us about ancient burials, but here was an ancient human who could tell us something about everyday life (I guess murder was part of that daily life too). Ötzi possessed leather shoes, a pack with food, and a copper-bladed axe. Further research showed that his gut was crowded with eggs of the parasitic whipworm *Trichuris trichiura*. Excitingly, he also carried leather straps with two cork-shaped fruiting bodies of the birch polypore fungus *Fomitopsis betulina* (formerly known as *Piptoporus betulinus*). This fungus makes its living off decaying birch wood, where it produces beautiful and smooth brown mushrooms.[2] It is also well-known to have antibiotic and laxative effects, suggesting that Ötzi was treating his infection.[3] Scientists posit that the toxic effects of the fungus would have killed the worms, and the laxative effects would have helped expel the worms from the gut. Until recently, Ötzi was known as the oldest evidence for medicine in humans.

Since those hikers found Ötzi, scientists have discovered that Neanderthals, those close but extinct relatives of modern humans, had already been using medicine tens of thousands of years before Ötzi's time. In a study published in 2012, Karen Hardy of the Universitat Autònoma de Barcelona and colleagues investigated calculus on the fossilized teeth of five Neanderthals from the El Sidrón cave in the north of Spain, where Neanderthals lived between 47,300 and 50,600 years ago.[4]

Calculus is the hardened mix of minerals from saliva and the bacteria and yeast cells that make up the plaque on teeth. In essence, it is a hardened fossilized graveyard of the microorganisms that inhabit the mouth and are responsible for tooth decay. Interestingly, during the formation of calculus, lots of things can get trapped in between the deposited layers, including smoke particles, tiny bits of food, and chemicals. Using a combination of high-resolution microscopy and chemical analysis, Hardy and colleagues found that one individual carried chemicals from yarrow (*Achillea millefolium*) and chamomile (*Matricaria chamomilla*). Neither plant is used as food by people, but both are used as medicine by many cultures around the world: yarrow as an astringent, and chamomile for stomach complaints and nervousness. The conclusion: Neanderthals most likely used these plants as medicine as well. (Why do the plants contain these medicating compounds in the first place? Don't worry, we're getting to that.)

Given that animals, ranging from ants to apes, use medicine, it is not at all surprising that we can find evidence for ancient use of medicine in humans. But what these archaeological studies show us is that people and other animals have a lot in common when it comes to medicine. Of course, nonhuman animals do not practice medicine to the same extent that humans do. They do not perform brain surgery, treat cancer with radiation, run genetic tests, or develop vaccines against pandemic viruses. But they use plants, fungi, microbes, or even animals that are loaded with toxic chemicals. In other words: both animals and humans use drugs. I want to use this chapter to discuss why these toxic chemicals are so readily available in nature. But before I do that, let me explain what I mean with the word "drug."

According to the Merriam-Webster online dictionary, one definition of medicine is "a substance or preparation used in treating disease." I will refer to such substances and preparations as "drugs." In some languages (such as Dutch, my native tongue), the word "drug" is often associated with addictive substances that one buys illegally on the street, not with medicines that are prescribed by doctors to treat infections or heart conditions. However, I will use the word "drug" in its more generic form, to describe all those substances that humans and animals use to treat their infections and other afflictions. This is justified knowing the origin of the word: "drug" derives from the French word *drogue*, referring to dried herbs.[5] This makes sense because for much of human history, plants have been a major source of drugs.

For some of us, it may seem that drugs are far removed from nature. We may go to a doctor in a brightly lit building that smells of antiseptics. We may be examined with modern technological devices. And after doctors prescribe a drug for our ailment, we may run to the pharmacy, where we stand in line to receive a vial or box with perfectly shaped, high-tech pills with unpronounceable names. Clearly, those drugs we receive are the product of human scientific ingenuity and technological advancement. Certainly, during the last 150 years, advanced science and technology have created completely novel chemicals; however, pretty much all drugs before that—and even the majority of drugs since then—have come from natural sources, especially plants.[6]

Take the pain reliever morphine, which is produced by opium poppies (*Papaver somniferum*), a plant that grows in many people's gardens. People have been using opium for over four thousand years.[7] It became widely used by the ancient Greeks and Romans. In the seventeenth century, opium—mixed with whisky or rum—was a common medication used during surgery. In the early 1800s, opium had become a social drug, and US citizens could buy it at local general stores.[8]

Quinine is an antimalarial drug that occurs in the bark of the Cinchona tree. The Quechua people in Peru used the bark long before French apothecaries isolated the chemical in 1820. Another widely used antimalarial drug is artemisinin, which comes from sweet wormwood (*Artemisia annua*). This drug was discovered when the Chinese government set up a research program to discover antimalarial drugs from traditional Chinese medicine.[9]

The anti-cancer drug Taxol was also discovered through screening plant extracts. Researchers found it in the bark of the Pacific yew tree (*Taxus brevifolia*) after the National Cancer Institute in the United States initiated a program to discover new drugs by screening large libraries of natural products and

FIGURE 7.1. Poppy head with opium latex. Photo by Lukas Puchrik / Shutterstock.

extracts.[10] Digoxin is a chemical that was first isolated from purple foxglove (*Digitalis purpurea*) in 1930; it is commonly used to treat abnormal heart rhythms and heart failure. Cocaine comes from coca plants and causes stamina and euphoria in humans. And as we have seen before, even the wonder drug aspirin finds its roots in plants, specifically in the bark of willow trees.

I bring up these examples to demonstrate that many well-known drugs are based on plants and that people have long recognized the healing power of these herbs. But plants are not the only organisms that contain chemicals that act as drugs. Fungi are another great source of drugs, as demonstrated by the mushrooms that Ötzi the Iceman carried, and by the first mass-produced antibiotic, penicillin, which is produced by a mold. And even animals have been sources of human drugs. The magical cone (*Conus magus*) is a predatory marine snail that uses venom to paralyze or kill fish; scientists isolated a protein from this venom and developed it into a drug to treat chronic pain.[11] And one treatment of type 2 diabetes mellitus is based on the drug exenatide, which was derived from a protein in the saliva of venomous lizards called gila monsters (*Heloderma suspectum*).[12]

While it is clear that many organisms are great drug sources, it is less clear why these organisms contain drugs in the first place. Take plants, which have been around for hundreds of millions of years. Surely, they did not evolve drugs for the benefit of modern humans, which have only been around for some two hundred thousand years? Surely, poppies did not start producing opium so humans could relieve pain? And surely, wormwood did not create artemisinin to fight malaria-causing parasites in the human blood stream? Quite right. It just so happens that these chemicals are hugely beneficial for humans. (Note, though, that despite their benefits, these drugs can also cause addiction and kill people; as so aptly described in Noah Whiteman's book *Most Delicious Poison*, there is a fine line between experiencing the benefits of drugs and their harmful effects through addiction.[13]) To understand why they evolved, we'll begin by looking at the struggles that plants experience in the face of millions of herbivores and pathogens that try to eat them. We can then use the insights from plants to better understand why fungi, some animals, and other organisms also contain chemicals that act as drugs.

VERY HUNGRY CATERPILLARS

Scientists estimate that there are about four hundred thousand species of plant on our planet. While this seems like a lot, they are far outnumbered by the insects, of which there are likely more than 5 million. Many of these insects eat plants. And they are not alone: plants also form the basic diet for most other animals, including giraffes, goats, and elephants. And then there are pathogens, such as mosaic viruses and rust fungi, that consume plant tissues. We may think of plants as crucial food to all these organisms. We may even argue that provision of food is the reason for plants' existence. But we would be wrong. Evolution does not work that way. Organisms do not evolve for the good of other species. And plants certainly do not think of themselves as mere food for animals. If we were to anthropomorphize plants, we would say that plants do not want to be eaten and that they work very hard to protect themselves against their enemies.

At this point it would serve us well to remember one very important thing about plants: they cannot walk. When a zebra gets chased by a lion, it can run away. But when a clover gets approached by a rabbit, it cannot bolt. And so, plants have evolved a staggering arsenal of defenses that help them survive in the face of millions of hungry herbivores, parasites, and pathogens. Milkweeds ooze generous quantities of milky latex that glues together the mouthparts of monarch caterpillars;[14] willow trees have tough leaves that damage the mandibles of beetles feeding on them;[15] grasses incorporate silica in their leaves to break grasshopper mouths;[16] hollies sprout nasty spikes on their leaves that skewer tent caterpillars;[17] acacias are covered in thorns that hurt the tongues and lips of kudus;[18] pea plants have waxy leaves that repel weevils (beetles);[19] and

Carolina horse nettles (*Solanum carolinense*) have hairs on their leaves that poke holes in the guts of tobacco hornworms.[20] Beyond these mostly mechanical defenses, plants are true masters at what can best be called chemical warfare.[21]

Chemicals come in many varieties, and many of them are involved in the basic metabolic processes that keep plants (and other organisms) alive. These include carbohydrates (sugars), fats, and amino acids (the building blocks of proteins). Scientists call these chemicals primary metabolites. Plants also contain a vast library of other chemicals with fancy names such as terpenes, alkaloids, saponins, quinones, and flavonoids. These chemicals have been known for hundreds of years, but because they do not directly support the basic processes of carbohydrate, fat, and protein metabolism, scientists long thought of them as waste products that have no specific function. As such, they coined them secondary metabolites.

Now we know, however, that many of these metabolites are crucial for plant survival. Plants use good-smelling volatiles and good-looking pigments to attract pollinators, or to protect themselves from harmful ultraviolet radiation. Some chemicals are used to kill other plants, and the function of others is to kill bacteria, viruses, and fungi. Yet others are there to provide protection against hungry mouths. Indeed, the evolution of chemical defenses is a major reason for the success of the flowering plants and their radiation into hundreds of thousands of different species since their origin in the Cretaceous, some 130 million years ago.[22]

The study of plant secondary metabolites took off in 1806 when the German pharmacist Friedrich Wilhelm Sertürner isolated morphine from the latex of the opium poppy. His demonstration that the activity of a drug can be attributable to a single chemical spurred the widespread study of natural products.[23]

This led to the discovery of all sorts of compounds. And although it was clear that many of these compounds were toxic or bitter-tasting, scientists maintained that these chemicals were waste products until the 1950s. In their mind, any useful function of these chemicals for plant life was simply accidental.

That view changed not because of open-minded chemists or plant biologists, but because entomologists—scientists who study insects and related animals—became interested in the role of these chemicals. Plants have been attacked by insects for hundreds of millions of years. Yet, plants are still here. And that means they must be quite good at protecting themselves. It was Gottfried Fraenkel, a University of Illinois professor, who wrote a landmark paper in 1959, in which he concluded that secondary metabolites are there for one major reason: to protect plants from insects.[24] Fraenkel explained that many plant chemicals are toxic and repellent to insects in general. But he also described how these same chemicals act as attractants to small numbers of insects. For example, diamondback moths only feed on plants in the crucifer (cabbage) family, which contain glucosides. Similarly, while essential oils in umbellifers (plants such as parsley, carrot, and celery) are toxic to most insects, the butterfly *Papilio ajax* uses them to recognize these plants as a suitable host. As such, the story that started unfolding in Fraenkel's mind is that the onslaught of plants by insects facilitated the evolution of a wide variety of protective chemicals in plants—and that some insects would evolve countermeasures and specifically use those protective chemicals to locate suitable food plants.

Fraenkel's paper was in fact a rediscovery of findings by Christian Ernst Stahl of the University of Jena, in Germany, who concluded in 1888 that plant chemicals were there to protect against herbivory by snails.[25] Like Darwin and Wallace, it was another case of two scientists looking at the world and coming to the same conclusion.

In 1964, Paul Ehrlich and Peter Raven at Stanford University took Stahl's and Fraenkel's work to a new level. Specifically focusing on butterflies, they showed that different species of butterflies use different species of plants to lay their eggs and provide food for their baby caterpillars. Ehrlich and Raven suggested that the wide variety of flowering plants that evolved since the Cretaceous arose because different plants evolved different chemicals to protect themselves. And as they did so, insects evolved countermeasures, which then required plants to diversify even more and to up their defenses. This would put more pressure on insects to overcome these stronger defenses, and so the species would keep on responding to each other in a coevolutionary arms race, much like we saw for hosts and parasites and pathogens in chapter 3. As such, Ehrlich and Raven suggested that plant diversification led to insect diversification—which then led to additional chemical defenses. And as specialized insects became more resistant, plants became more toxic.

CHEMICAL WARFARE

Scientists now broadly accept that the coevolutionary process between plants and insects has spurred the evolution of many secondary chemicals. But not all. Insects are not the only reason for the existence of these compounds. Plants also use them to protect themselves against mammals that try to eat them; bacteria, fungi, and viruses that try to infect them; and other plants that compete with them for sunlight, nutrients, and space. Paradoxically, even though plants evolved these chemicals to cause harm to their enemies, secondary chemicals often have beneficial effects for humans and other animals as well.

Take morphine, quinine, and cocaine. These chemicals belong to a class of compounds known as alkaloids.[26] Scientists have found over twelve thousand alkaloids in about three hundred different plant families. Alkaloids taste bitter, which deters herbivores. Many alkaloids are toxic to insects and other animals, while others act as antioxidants: molecules that trap oxygen molecules possessing excess and damaging energy, which are produced as byproducts of cellular metabolism. Capsaicin is an alkaloid that makes chili peppers spicy and kills velvet army worms (*Spodoptera latifascia*) that try to eat pepper plants.[27] Nicotine, used by finches and sparrows to fumigate their nests, is the mind-stimulating and addictive ingredient in cigars, pipes, and cigarettes. It is produced by many species of tobacco in the genus *Nicotiana* and messes up the nervous system of insects, leading to paralysis and death.[28]

Caffeine, the alkaloid in coffee and tea that keeps billions of people alert and awake during boring meetings, kills cabbage white caterpillars, snails, fungal pathogens, and microbes.[29] As for cocaine, this alkaloid messes up the nervous system of insects, causing tremors and death;[30] the stimulation and euphoria experienced by humans is an accidental side effect of the millions of years of evolution that led to the production of cocaine by coca plants to protect themselves against hungry insects. The morphine produced by poppies most likely evolved as a neurotoxin to kill insects; as a side effect it acts on human nerve cell receptors to relieve the sensation of pain.[31]

The antimalarial artemisinin and the anti-cancer drug Taxol are terpenes, another class of around thirty thousand chemicals. Plants use terpenes to send airborne messages to insects, to produce hormones, to provide colored pigments, and to kill insects and microbes. Plants leak artemisinin into the soil to kill nearby plants and relieve competition for space, light, and nutrients,[32] while Taxol's natural role in the yew tree is most likely to kill insects.[33]

The secondary chemicals that plants produce are not always there to hurt, poison, or kill herbivores directly. In some cases, the chemicals are used to kill part of the plant itself—and with it the herbivore. When cabbage white butterflies (*Pieris brassicae*) lay eggs on black mustard plants (*Brassica nigra*), the plant responds to the egg by forming necrotic tissue, which kills the egg or makes it fall to the ground.[34] Although the plant loses part of a leaf, that is preferable to being eaten by a hungry caterpillar. Salicylic acid, the chemical that formed the basis for aspirin, comes from a similar process: when infected with a virus, plants use salicylic acid to set in motion a process of necrosis that stops the virus from spreading through the plant.[35]

Of course, plants are not the only organisms that produce such chemicals. Fungi and even animals have developed similar chemicals in the same kind of evolutionary arms races as the one between plants and their enemies. As we saw at the start of this chapter, Ötzi the Iceman used polypore mushrooms, which contain chemicals called triterpenes, as medicine. One of the most widely used antibiotics of all time, penicillin, was likewise obtained from a fungus. And as we saw in chapter 4, animals may also use medicinal compounds produced by other animals. Capuchin monkeys and lemurs use toxic millipedes to repel mosquitoes. Away from land, corals and sponges produce many antimicrobial compounds. And spiny dogfish, a species of shark, produce an antibiotic known as squalamine, which kills bacteria, fungi, and protozoans.[36] In all of these cases, the toxic chemicals provide one important function: protection from natural enemies, including predators, parasites, and pathogens.

FIGURE 7.2. A lemur bites a millipede, which then releases insect-repellent chemicals. Photo by John Downer / Nature Picture Library.

FOOD AND MEDICINE

Altogether, plants, fungi, and other organisms provide a vast pharmacy that humans and other animals can use to find drugs to battle their parasites and pathogens. In the next chapter, we'll explore how animals make use of that pharmacy and know how to navigate this myriad of drugs. Before doing that, I want to make one more important point: not all animal medication hinges on the use of toxic chemicals. Animals can use food as medicine, too.

I have mentioned before that defining animal medication is difficult and that scientists use different definitions. Having said that, most scientists argue that medication involves chemicals or substances that are toxic. This means that most scientists would discount the use of food as medicine. They would count the use of plant secondary chemicals as medication but not the use of the primary chemicals that include proteins, fats, and carbohydrates. But not everyone agrees. Indeed, during the insect medication symposium in Finland that I attended in 2022, and which I mentioned in the previous chapter, Sheena Cotter of the University of Lincoln in England made the case that animals can also medicate themselves with nutrients. Reporting on her own work, she explained that African armyworm caterpillars (*Spodoptera exempta*) are much more likely to survive infection with bacteria and viruses when they consume high-protein diets.

"And caterpillars know this," she added, explaining that when infected, they immediately shift their diet choices from high carbohydrates to high protein.[37]

Cotter went on to summarize other research on nutrients. For carnivorous burying beetles (*Nicrophorus vespilloides*), high fat diets are best to survive infection with virulent bacteria.[38] In contrast, for crickets, low-fat diets are better for survival when infected with bacteria, and when given a choice, crickets that are ramping up their immune responses against bacterial pathogens avoid fat as much as they can.[39]

In many cases, diet effects on infection occur because of changes in immunity, with high protein levels helping mount stronger immune responses to parasites and pathogens.[40] For example, sheep infected with worms (*Trichostrongylus colubriformis*) increase their dietary intake of protein, which can boost their immune system.[41] Likewise, Egyptian cotton leafworm caterpillars (*Spodoptera littoralis*) that eat diets with high ratios of protein to carbohydrates have stronger immune responses and greater resistance against body-liquifying viruses.[42] And African armyworms that eat high-protein diets have greater immunity and resistance against bacterial infections.[43]

But food can also reduce infection in other ways. In one study, Cotter found that while African cotton leafworm caterpillars infected with the bacterium *Xenorhabdus nematophila* survive better on a high-protein diet, they actually mounted *weaker* immune responses than on a low-protein diet.[44] Instead, she found that the high protein levels in beneficial diets increased the protein concentration of the insect's blood, which made the blood less hospitable for the bacterium.[45]

Perhaps it is not surprising that animals can use food as medicine. Much of traditional human medicine is based on food as well. A major focus of traditional Chinese medicine is holistic healing and maintaining a healthy lifestyle, which includes drinking green tea. Similarly, a major component of Ayurvedic medicine is to combine a healthy diet with herbal treatments and yoga. Even Western medicine, now mostly characterized by drugs and surgery, is rooted in a focus on holistic health that includes a healthy diet.[46]

Western medicine took a great leap during the Classical period of Ancient Greece, the time best known for Hippocrates and his followers. The philosophy of this period was that disease should no longer be viewed as a supernatural phenomenon and that rationality, naturalism, and science should be at the root of medicine. Hippocratic medicine remained holistic in nature, with healing focused on treating the entire patient and not just the disease or a single body part. Hippocrates is often quoted as saying: "Let food be thy medicine and medicine be thy food." (Though it must be said that this phrase doesn't actually appear in the Hippocratic Corpus, and instead appears to be a recent fabrication to suggest that Hippocrates equated food with medicine.[47]) While the exact quote is not quite correct, it is true that Hippocratic medicine was heavily based on diet. In terms of the order of medical interventions, altering diet came before the use of drugs and surgery. The Hippocratic Oath, which doctors still take today to express that they will do no harm to the patient, includes, "I will apply dietetic measures for the benefit of the sick according to my ability and judgment; I will keep them from harm and injustice." Thus, diet became a moral value and responsibility in medical practice.[48]

Separating nutrition from medicine is often difficult. And the difference between food and drugs can be fuzzy. It was Swiss physician and alchemist Paracelsus who said that "solely the dose determines that a thing is not a poison."[49] Paracelsus is often viewed as the father of toxicology. His contemporaries included Leonardo da Vinci, Martin Luther, and Nicholas Copernicus, and his statement that the dose makes the poison is well-illustrated by coffee. Small amounts of coffee act as a mental stimulant; high amounts can kill. That bitter-tasting alkaloid caffeine can thus serve as drug and poison.

In some ways the fuzzy boundaries between food and medicine, and between drug and poison, make animals' jobs to medicate themselves and each other harder. They cannot simply separate supermarket from pharmacy. They cannot easily go from a food aisle to a drug aisle in a megastore. Instead, they are faced with an endless choice of nameless plants, fungi, animals, and other items that could feed, cure, or kill them. So, how do they navigate this immense food-drug jungle? How do they make the choices that keep them healthy and alive? As we will see in the next chapter, some of these choices are innate, while others have to be learned through trial, error, and social interactions.

8

Living and Learning

Whenever I talk about my research on monarch butterflies, whether at a butterfly festival, during a scientific presentation, or at an interview for the local radio station, I always get the same question. "How do they know?" How do monarchs know what plants to use to medicate their offspring? It is a great question, and the answer is disappointing. I simply don't know. To be honest, I am not even sure how monarchs can tell that they are sick. But I usually begin my answer by speculating that the whole physiology of infected monarchs is so disrupted that they must feel different, and probably don't feel good. And that malaise may alter their preferences and aversions for some milkweeds, so that they pick the more toxic and stronger-smelling and -tasting milkweeds when infected.

Luckily, we know much more about a different species of caterpillar. In chapter 6, we met Michael Singer, who showed us that woolly bear caterpillars eat plants rich in pyrrolizidine alkaloids to kill the parasitoid flies that eat them from the inside out. As I was talking with Singer about these caterpillars, he explained to me that infected woolly bears develop a preference for medicinal food because of changes in their taste sensation. Like most other animals, caterpillars have cells with receptors that allow them to taste their food. Humans have taste buds with cells that have receptors that can distinguish between five different flavors: salty, sweet, bitter, sour, and umami (savory). In contrast, woolly bear caterpillars have four different cells, with each responding to a different range of chemicals.

"The most intriguing thing is that one of these four cells only responds to pyrrolizidine alkaloids," says Singer.[1]

After Singer finished his PhD research, he continued his research collaboration with Liz Bernays. Together they went on a quest to figure out how infection with parasitoid flies changes the taste receptors of the woolly bear caterpillars. Using some very fine tools and steady hands, Bernays managed to insert tiny electrodes into the taste buds of woolly bears. This allowed her to measure the firing rates of the cells: higher firing rates mean greater taste responses. She immobilized caterpillars in small glass vials and placed small droplets of alkaloid solution on their mouthparts. The result: infected caterpillars had much greater firing rates than uninfected caterpillars, showing that infection increases their taste for pyrrolizidine alkaloids.[2]

What this means is that caterpillars infected with fly maggots develop a stronger taste for alkaloids, which makes them seek out plants with higher concentrations of these chemicals. And as a result, they end up medicating themselves. Interestingly, the response is purely innate and physiological. The caterpillars do not need to know they are infected. They do not need to learn from experience. They simply listen to their bodies, which tell them to eat more of those highly toxic chemicals when they are infected with those horrendous maggots. In some way, the caterpillars' self-medication is hardwired in their bodies and passed from generation to generation. Over the long course of evolution, it was the individual woolly bears able to respond to infection by increasing their consumption of alkaloids that survived and reproduced. Thus, the genes that enabled them to change their sense of taste as a response to infection were passed on to their offspring.

It is quite likely that the medication behaviors in other insects are similarly driven by innate mechanisms. You may remember the African armyworm caterpillars that shift their diet to obtain more protein. The most likely mechanism is that their bodies somehow respond to the quick replication of specific viruses that start lining their guts within twenty-four hours of infection. And insects are likely not alone in this: as we will see later, cats, dogs, and other mammals also seem to have innate responses to medicinal compounds that make them feel better.

But innate responses are not the only way by which animals develop their medication behaviors. As I discuss in this chapter, many animals make associations between the use of different foods and how they feel, and learn from those experiences. And when such individual learning is not enough, animals can also obtain their medical knowledge from others.

SEXY GENES AND BITTER PILLS

The fact that animal medication can be an innate behavior, driven by an animal's genes, has profound implications. Until recently, there was quite some skepticism among scientists about animal medication, with many believing that only animals with large brains and the capacity to learn over time could develop medication behaviors. Indeed, my colleagues and I had to remove a section from one of our research papers that hypothesized that monarch butterflies could use milkweeds as medication because one of our reviewers thought there was simply no way that monarchs could do this. But if medication behaviors are innate, passed down through genes, it means that many more animals than originally thought can use medication: ants in addition to apes; bees in addition to bears; and caterpillars in addition to capuchins.

I think that the skepticism about "simple" animals being able to medicate stems from the general unease that many scientists have felt about the idea that behaviors can be controlled by genes. Indeed, I believe it goes back to the nature versus nurture debate, which is all about whether behaviors are determined by genes or by the environment. For a long time, behaviors were seen as the product of environment and learning, to be contrasted with physical traits such as eye color and blood type, which were deemed to be shaped by genes. However, it is now abundantly clear that for most of these traits, both genes and environment play a role. As evolutionary biologist Marlene Zuk explains in her book *Dancing Cockatoos and the Dead Man Test*,[3] the dichotomy between genes and environment is not helpful, and every behavior is the result of both genes and environment. Ultimately, what this means is that any behavior has some genetic component to it. Because genes are the material that evolution shapes over time, it follows that animal behaviors evolve just like any other animal trait, such as hair color, tail length, and tooth number.

That behaviors are influenced by genes, and that they therefore can and do evolve, may not seem that shocking now, but it once was. It was not until the 1950s and 1960s that scientists started collecting conclusive evidence that behaviors are subject to genes. One oft-cited pioneer was Walter Rothenbuhler, of Ohio State University, who studied hygienic behavior in honey bees. After the queen bee has laid an egg in a honeycomb "cell," a larva will hatch from it and will be fed by worker bees. Next the worker bees will cap the cell with wax so that the larva can pupate. If any of these cells are infected with American foulbrood bacteria, honey bees uncap the brood cells and dump the sick and dying larvae outside the hive. By choosing bees from colonies that had this behavior and mating them with bees from colonies that did not, Rothenbuhler found four types of offspring bees: those that were fully hygienic, uncapping cells and removing dead larvae; those that uncapped cells but left the larvae to rot in their brood cells; those that did not uncap cells but removed dead larvae from cells that had been opened by researchers; and those that neither uncapped cells nor removed larvae from uncapped cells.[4] This segregation into four types of behavior told Rothenbuhler that the behavior is likely controlled by two genes: one that makes bees uncap a brood cell, and one that allows them to remove the larva. This simple relationship between gene and behavior was so important that it helped lay the foundation of Richard Dawkins's argument in his famous book *The Selfish Gene* that genes are responsible for all sorts of behavior, including aggression, cooperation, and altruism.[5]

Although Rothenbuhler's study was groundbreaking, it was not the first to show a direct link between genes and behavior. Back in 1956, another landmark paper was written by Margaret Bastock, a PhD student working with the Dutch ethologist Niko Tinbergen at the University of Oxford. Bastock investigated fruit flies. While normally brown, a single gene mutation makes flies yellow instead. And as it turns out, yellow males are less motivated to court females. Female fruit flies get turned on by a good male wing vibration, but yellow males vibrate their wings in less attractive, shorter bouts and at longer intervals than brown males, thus showing a direct link between a gene and courtship behavior.[6]

Since these papers by Bastock and Rothenbuhler, many studies have shown that behaviors are influenced by genes, and thus

FIGURE 8.1. In a process called hygienic behavior, honey bee workers uncap brood cells of infected larvae, remove the larvae, and then discard them outside the hive. Photo by Gianluigi Bigio.

can be passed from parents to offspring. These behaviors range from courtship frequency in junglefowl to preening in Japanese quail and learning in pigs.[7] One that I find especially fascinating is the sexual stroking behavior of spotted cucumber beetles.[8] As with many insects, mating in cucumber beetles is a multifaceted process, whereby penetration does not necessarily lead to sperm acceptance. Thus, when a male spotted cucumber beetle manages to penetrate the vagina of a female, he still needs to persuade her to accept his sperm by opening a duct that leads to her bursa copulatrix, an organ in which she stores sperm for future egg fertilization. To gain access to her bursa, a male needs to use his antennae to stroke the female's antennae, eyes, and front legs. And he needs to do it fast and furiously. In a fascinating series of experiments, researchers not only demonstrated that faster-stroking males had greater mating success, but also that they fathered fast-stroking and sexually successful males. In evolutionary circles this is known as the sexy son hypothesis, whereby females prefer to mate with those males that endow them with sons that will be sexually successful themselves. (The beauty of science is that it provides a wonderful series of lines to use during introductions at parties and official events. "Nice to meet you, what do you do for a living?" "Oh, I spend my days measuring whether fast-stroking male beetles are better sexual partners.")

As we have discovered increasing evidence that behaviors are heritable, and therefore can and do evolve, we have also learned that many animals that we may think of as very simple can do some spectacular things. This means that intricate behaviors do not require high cognitive processes. Instead, intricate behaviors can be driven by simple rules—or rules of thumb as Marlene Zuk likes to call them. In her own work, Zuk has long studied Pacific field crickets in Hawaii. Male field crickets rub their wings to produce the typical cricket song, which they use to attract females. The problem in Hawaii, however, is that parasitic flies eavesdrop on the crickets. Using the male crickets' efforts to land a female, the flies will locate the males, and leave tiny sticky larvae on and near them, which then invade the males' bodies and eat them from the inside out—similar to the flies that attack woolly bear caterpillars.

Astonishingly, Zuk has found that some male crickets evolved wings without the structures that produce sound when rubbed together. The problem for these males is that they can no longer attract females. But they have found a solution: by moving close to males that can sing, they can still encounter females and end up with a mate. Not that they know what they are doing, writes Zuk. They just have a simple rule that tells them: "If I do not hear a lot of crickets singing, I should move closer to the ones I do hear."

Zuk thinks that such simple rules underlie a lot of different behaviors. For example, when given a choice between red and green leaves, egg-laying cabbage white butterflies innately prefer green leaves. It is a simple rule of thumb that works because green leaves are common, and it does not require the butterfly to evaluate differences in color every single time it lays an egg.[9]

What is true for finding mates and laying eggs can be true for medication as well. Animals do not need to know that they are medicating themselves. Take monarch butterflies. They do not need to know that they are sick, nor that particular plants will cure their offspring. What matters is that when they are sick they change their preferences, that this behavioral change is heritable, and that these changed preferences make their offspring less sick. As a result, the offspring are more likely to survive and reproduce and pass on the ability to medicate to their own offspring.

These rules of thumb that "simple" animals like insects use may just as well apply to vertebrates with higher cognitive skills. Benjamin Hart, an emeritus professor of veterinary sciences and animal behavior, has hypothesized that the rules of thumb that animals use are based on specific tastes associated with medicinal properties.[10] Unlike woolly bear caterpillars, many animals may not have the ability to detect the exact chemicals that act against their parasites and pathogens. Indeed, much research has indicated that salt is about the only chemical that mammals can specifically taste, and that they increase its consumption when deficient.[11] But according to Hart, general rules may work just fine. His idea is that illness changes the palatability of food, and that it leads to reduced aversion for bitter, astringent, and repugnant plants. As many of us know, many medicines taste bitter; hence the expression of "swallowing a bitter pill." Although we may have aversions for strong bitter tastes, we also quite like to sample mildly bitter foods, such as coffee, tea, and chocolate.

Animals are much the same, and the sampling of bitter foods equips them well to learn what foods may have medicinal properties, allowing them to turn to those foods when sick. In one experiment, researchers provided laboratory mice with two types of drinking water: one plain and one supplemented with the bitter-tasting, antimalarial drug chloroquine. Both malaria-infected and uninfected mice readily sampled the chloroquine solution, and this reduced disease in the infected mice.[12]

As with human medicine, many of the medicinal plants used by animals taste bitter, including the milkweeds used by monarch butterflies and the *Vernonia* plants used by chimpanzees (the colloquial name of *Vernonia*, bitter leaf, says it all). It is also important to note that chimp diets generally contain very few bitter-tasting plants, suggesting that they indeed use bitterness as a marker for medicine.[13] As such, simply tasting a particular plant can go a long way in telling an animal whether the plant may be medicinal.

NAUSEA AND NEOPHOBIA

The fact that many animals use innate behaviors to medicate does not mean learning is unimportant. Indeed, in many animals, learning—either by the animal on its own or from other animals in social groups—is crucial for their medication behaviors. While researchers may have traditionally turned to apes or dogs to study the process of learning, we now know that bees are expert learners too.[14] They can distinguish human faces, recognize different flowers, and even learn to associate specific symbols with different numbers, a prerequisite for mathematics.[15]

Excitingly, a recent paper showed that even animals without brains can learn. Specifically, researchers from Germany and Denmark found that box jellyfish are experts at learning to avoid obstacles when swimming around in experimental arenas. Jellyfish do not have a centralized brain, suggesting that much simpler nervous systems can help animals learn from their experiences.[16]

One important way in which animals learn is through association, as first investigated by John Garcia and colleagues in the 1970s. In one study, Garcia injected rats with apomorphine hydrochloride, a substance that causes abrupt nausea, followed by rapid recovery.[17] Garcia fed the rats either milk or grape juice before or after injection with the nauseating chemical, and found that the rats avoided whichever drink they received before the injection but preferred the drink given during recuperation. Thus, even though the actual milk or grape juice are in no way responsible for causing or relieving nausea, the rats associated their flavors with the onset and resolution of nausea. Many studies on rats have demonstrated the development of these negative associations, but rats also develop preferences for tastes that coincide with recovery from illness, thus showing that positive associations can also be acquired.[18]

Scientists have carried out similar ingenious experiments with goats and sheep.[19] In some, researchers investigated sugar-deficient lambs, and fed them straw—a low quality food—that was either flavored with apple or maple. As the lambs were eating the straw, the researchers infused sugar directly into their stomach-like rumens. By infusing the sugar, the researchers made sure that the animals could not taste the sugar; yet their bodies experienced the sugar surge. Afterwards, they gave the lambs choices between the two types of flavored straw. Astonishingly, they preferred the flavor they had experienced while experiencing a sugar infusion. Similar studies with lambs that were deficient for specific minerals showed that sheep learn to prefer foods with the minerals they are deficient in. With all these examples, it is the animal's bodies that make the associations and do the learning. The animals do not necessarily know what they are doing; they simply respond to what their bodies are telling them. When a sheep is low on sugar and eats sugary food, its body will respond positively, and the sheep will associate that food with feeling good.

Associative learning is very common, and has been demonstrated in many different animals, including slugs, grasshoppers, garter snakes, Atlantic cod, and crows, to name a few.[20] The learning that occurs with food preferences or taste aversion does not require higher-order cognitive processes. This is clear because animals can make associations even when being anesthetized (and are thus not consciously aware of any taste), or when rewarding or nauseating solutions are directly infused into their stomachs. But it is also clear because humans have these associations and we cannot override such aversions with cognitive control: even when we know that a particular food did not cause a particular illness, we retain the aversion.

If you have ever had food poisoning, you may be familiar with the following scenario. Say you ate a meal with a particular hot chili sauce, and became violently ill within hours following that meal; you may start blaming that sauce for becoming sick. Even when you know you had been eating that same sauce for many days prior, and that there must have been something else in that meal that made you sick, your body will simply not allow your mind to override the aversion. What's more: every time you even smell that sauce, your body will tell you to avoid it at all costs.

Another example comes from treatment of cancer with chemotherapy and radiation. Because these treatments have very strong side effects, causing fatigue, overall malaise, nausea, and vomiting, patients often develop aversions to foods or beverages they consumed before experiencing the illness caused by treatment.[21] In one study, researchers found that some patients undergoing infusion chemotherapy for breast cancer even developed aversions to highly preferred foods such as chocolate, spaghetti, and apple juice.[22] Thus, even though they know that it is the cancer treatment that makes them sick, the patients cannot override the aversion that their bodies developed.

Preferences also matter when it comes to choosing potential medications. In order to tell if animals are actually choosing to consume specific foods for health purposes, we need to understand whether animals can actually taste the compounds that they need. Animals have specific hungers for salt and water, creating cravings when they are deficient. But as far as we know, animals do not necessarily have specific taste receptors for all the different medicines they use. As I discussed earlier, they could use some rules of thumb, choosing bitter or astringent plants.

Alternatively, animals may not have preferences for medicinal foods, but instead might be consuming medicinal foods as an alternative because they have developed an aversion for familiar foods.[23] Experiments by Paul Rozin at the University of Pennsylvania showed that rats deficient in thiamine (also known as vitamin B1) preferred to eat new foods, and afterwards avoided the familiar thiamine-deficient food, even when it was the only food available to them.[24] This suggests that deficient animals feel unwell, and that they associate their familiar food, lacking in important nutrients, with illness. They are literally sick of their old food and willing to try new things.

One important thing these studies have taught us is that animals that feel sick or deficient are more likely to seek out novel diets. When it comes to eating, most (healthy) animals are very conservative with their food choices, since making a wrong choice can be deadly. As we have seen numerous times, most plants and many other organisms are toxic, ready to kill or otherwise harm the consumer that dares to eat them. This is why it is difficult to use poison to treat rat infestations. Simply sprinkling poisonous food around does not work, as rats avoid it. Pest control experts will tell you that one solution is to provide novel food to rats, and once they have sampled it and incorporated it into their diet without any negative associations, add the poison.

Scientists describe the aversion for new foods (or other things) neophobia, or "fear of novelty." Neophobia is a great way for animals to protect themselves against poisoning. By very carefully trying new foods, animals prime their bodies to make associations between the taste of the new food and any ill effects they may experience from eating it.[25] Over the last decades, scientists have concluded that sickness and deficiency make animals lose some of that neophobia. As a result, the animal is more eager to try new foods. And by trying new foods, animals can then associate those new foods with improved feelings.

Some scientists have questioned the ability of animals to medicate by arguing that the consumption of a particular food and the recovery from symptoms is not immediate enough, and that there is no way that animals will be able to associate the consumption of something with a change in condition many hours later. However, research has shown that the associations between improved health and particular foods do not have to be immediate: animals may learn associations even when the consequences of eating particular foods are experienced as late as twelve or twenty-four hours later.[26] For example, when given a choice between plain tap water and sugar water, rats preferentially drink the sugar water. But when injecting rats with a nauseating drug twenty-four hours after consumption of the sugar solution, researchers found that the animals developed an aversion of sugar and subsequently chose plain tap water over the sugary drink.[27] Thus, animals are very capable of associating the onset or relief of symptoms with consumption of particular foods.

Overcoming neophobia is especially important in order for animals to eat foods with high concentrations of secondary chemicals. Medicinal foods often taste bitter and are generally unpalatable, and many animals avoid them when they do not need them (although, as I mentioned, animals may sample bitter foods, which allows them to get acquainted with their natural pharmacy). Scientists believe that the major disruption in animal health due to infection can push animals over the aversion that normally stops them from eating toxic foods. In her book *Seeds of Hope*, Jane Goodall relates how she would sometimes lace bananas with the bitter-tasting antibiotic tetracycline and then offer them to chimpanzees with infected wounds.[28]

"What amazed me," she writes, "was that whereas a sick chimpanzee accepted and ate the fruit without hesitation, once he or she was better, such bananas were rejected after a cursory sniff."

Similarly, when foraging in pastures, lambs with high burdens of parasitic worms eat more diverse diets than those without worms and are more likely to sample bitter-tasting medicinal plants.[29]

YOU ARE NOT ALONE

To develop effective medication, many animals benefit from social interactions with other animals. One way in which social interactions facilitate the development of medication behaviors is that animals are more likely to lose their neophobia when surrounded by others. This may sound familiar for those of us traveling to new countries. I remember living in the Malaysian state of Sabah for four months, where I was much more willing to try unfamiliar foods (such as soup made with pig intestines) when surrounded by colleagues than when going for dinner on my own (when eating by myself, I would opt for the chicken fried rice I had grown up with in the Netherlands).

Other apes are much like us. One study on chimpanzees, gorillas, and orangutans in zoos, for example, showed that these apes quickly lost their neophobia for medicinal plants when living in groups.[30] And the same is true for sheep: lambs that live with their mothers are more likely to try new foods than those living in individual pens.[31] But animals may also benefit from social interactions by conforming to what others are doing or because they can directly learn medication from role models.

Conformism is common in social animals, and many different behaviors, ranging from the learning of songs in birds to the use of tools in chimpanzees, stem from the urge to belong and fit in.[32] With respect to medication, studies have shown that animals tend to eat foods that other group members eat—even when it goes against their own initial preferences and aversions. In one study, researchers provided rats with two different foods, one flavored with cinnamon and one with cacao[33]. They then paired the consumption of cinnamon-flavored food with the injection of lithium chloride, which makes rats feel sick. As a result, the rats associated the cinnamon-flavored food with nausea and started avoiding the food. The researchers then introduced "demonstrator" rats into the cages, who had been conditioned to prefer cinnamon-flavored food. As these rats were chomping away at the cinnamon food, the experimental cinnamon-averse rats overcame their aversion and started to eat the food once again. Similarly, when experimental rats developed preferences for cinnamon-flavored food supplemented with sugar they abandoned that preference when confronted with demonstrator rats that preferred to eat cacao-flavored food. In both cases, rats conformed to social influences.

Researchers studying spotted hyenas found similar results: when dominant females were offered a novel food with lithium chloride, they developed strong aversions for those foods. But when surrounded by the other hyenas in their group—who had not developed an aversion—the dominant females overcame their aversion. This shows that even dominant animals are subject to social conformity.[34]

When it comes to learning from role models, many animals learn what foods to eat by copying parents or other individuals in their social groups. In one study, experiments showed that meerkat pups were more likely to eat hard-boiled eggs when they observed other meerkats eating them, and also learned more quickly that scorpions are meerkat-appropriate food items.[35]

Elephants learn what to eat by touching each other's trunks and mouths and even sticking their trunks into the mouths of other elephants to sample the food they eat.[36] And chicks learn from hens what food items to peck at before they can develop positive associations between ingesting particular foods and gastrointestinal feedback. (This also suggests that the randomly pecking rooster sidekick in the Disney movie *Moana* either lost

FIGURE 8.2. Michael A. Huffman at age twenty-one, studying Japanese monkeys at Arashiyama, near Kyoto. His studies of stone handling in these monkeys showed that primates develop cultural behaviors through social conformity and learning. Photo owned by Michael A. Huffman, taken by Michio Nagamune.

his mother at an early age or simply did not pay attention to her lessons).[37]

Chimpanzees are highly social animals, and juveniles pay close attention when adults eat unusual food items.[38] Studies have shown that individuals develop self-medication by copying each other. You may remember how Michael A. Huffman and Mohamedi Seifu Kalunde observed how the chimp Chausiku collected *Vernonia* shoots and sucked the bitter pith to remove intestinal parasites. As she was breaking off the branches and biting the pith into small pieces, she would drop some onto the forest floor. As she went about her business, her infant Chopin placed some of the discarded pieces in his mouth, then spat them out quickly, presumably because of its bitterness.[39] By copying her behavior, Chopin would have learned to associate this particular plant with a bitter taste.

Primatologists have also witnessed that infants copy the leaf swallowing that many adult chimpanzees do to dislodge parasitic worms from their guts during the wet season. This social learning of leaf swallowing not only explains why leaf swallowing is so common in wild chimpanzee populations, but also why different groups use slightly different methods. During one of my conversations with Huffman, he showed me both types, using a random leaf from a potted plant in his office and his own mouth and hands. With partial leaf swallowing, chimpanzees add the leaf sideways to the mouth, tear off a piece, then rotate it in the mouth before folding and swallowing it; with complete leaf swallowing, chimpanzees put the whole leaf in their mouth, pull it in with their tongue and lips to fold it multiple times over before swallowing it. Both types of leaf swallowing occur in nature, and together with colleagues in Italy, Huffman showed that both can develop in captivity as well.[40] The researchers introduced shoots of *Helianthus tuberosus* into two chimpanzee enclosures at the Parco Natura Viva–Garda Zoological Park in Verona, Italy.

H. tuberosus is a common plant in Verona; it has leaves with short, stiff hairs that look and feel like the leaves of the more than forty plant species used by chimpanzees, bonobos, and gorillas in Africa and gibbons in Thailand.[41] In one population of six chimpanzees, referred to as the Island Population, the alpha male rejected the leaves. But as soon as one female started swallowing the leaves, other chimpanzees followed her example. In the Tree group, the dominant male swallowed the leaves and the other individuals followed suit. Interestingly, the Island chimps developed partial leaf swallowing, while the Tree chimps developed complete leaf swallowing. As such, the researchers directly observed the development of different cultures by the different groups of chimpanzees. While both methods of leaf swallowing are effective, each group settled on a different method, simply because the first chimp to engage in leaf swallowing chose a particular method, and the other chimpanzees copied the behavior. Both these variants are used by wild populations of chimpanzees.

None of the chimpanzees in the zoo in Verona had had an opportunity to learn the behavior before, none of them was infected with worms, and none of them had encountered rough leaves before. Yet, in each group one chimpanzee spontaneously started swallowing the leaves and others followed suit. To Huffman, this also answers a long-standing puzzle in the field of animal medication: How is the behavior initially acquired?[42] Back in 1994, Huffman cowrote a chapter with Richard Wrangham in a book on chimpanzee culture. Together, the primatologists hypothesized that self-medication behaviors start as an innate response: this means that chimpanzees may be prone to swallow leaves simply because of the texture of the leaves. In this context it is important to realize that chimpanzees, bonobos, and gorillas all use the same species of plants for leaf swallowing, all with the same rough leaf surfaces. They never do it with smooth leaves.[43]

Although the swallowing of leaves may be innate, chimpanzees may be more likely to try it when they feel sick: again, feeling bad reduces animals' neophobia. And once the apes start swallowing leaves, they will learn to associate the behavior with relief from intestinal worm infections. If this sounds like an unlikely sequence, remember this: not all animals have to develop the behavior on their own. They can copy the behavior of others, then form associations between the behavior and disease relief, which is reinforced by observing many other individuals in the social group medicating themselves as well.

As Huffman told me: "When you look at the population of chimpanzees in Mahale, they know the plant is used by the group. Infants observe their mothers using the plant. They learn what plant, and how it is eaten. But they are not getting the whole information. The mothers cannot say: take this and you will feel better. So, there is individual learning. They will learn the benefits from experiencing it."[44]

As the leaf-swallowing apes so aptly show, animals can acquire medication skills through a combination of innate abilities, associative learning, and social interactions. How important each of these mechanisms is will vary from species to species. But what all animals have in common is that they do not need to be aware of how and why they are medicating themselves. There is a big difference between function (what a behavior is for) and motivation (why an animal performs the behavior). In his book *The Ape and the Sushi Master*, primatologist Frans de Waal explains this difference with respect to the healthy appetite of teenagers and pregnant women.[45] While the function is growth of the teenager or the baby, neither the teenager nor the pregnant woman consciously eats a lot to grow. Instead, their motivation is hunger. Similarly, he describes the case of a stallion fighting against other stallions to obtain a harem of mares. While the function of this behavior is to ensure procreation, its motivation is not.

And so it seems to be with animal medication. While the function of leaf swallowing is the expulsion of intestinal parasites, this is not the motivation for the development of the behavior. As the captive studies show, chimpanzees apparently swallow leaves because they are innately primed to do so and because they are a social species that tends to conform and imitate. Once engaging in the behavior, the chimpanzees start associating swallowing leaves with recovery from the symptoms of worm infection, which reinforces the behavior.

While I ended this chapter with chimpanzees, explaining how they use individual and social learning to develop self-medication, I want to reiterate that many animals use *innate* abilities instead. By the time a woolly bear has eaten enough alkaloids to kill parasitoid flies, it is turning into a pupa, never to eat plants with alkaloids again. And monarch butterflies do not even medicate themselves. Mothers medicate their offspring instead. This means that any benefit of laying eggs on medicinal milkweeds is experienced by the caterpillars, not by the mother, who therefore cannot learn from her own behavior.

That animals can use innate abilities to develop medication behaviors means that any animal, in principle, has the ability to heal itself or others. The implication is profound: although we now know of numerous animals that use medication, the actual number must be vast. As such, I have no doubt that the next few decades will reveal a whole collection of newly discovered animal doctors. And this is not just interesting to academic scientists. As we will see in the next few chapters, understanding and embracing these behaviors may be essential to developing better agricultural practices and improving animal welfare.

9

Woolly Wisdom

Many people think that domesticated animals are rather less sophisticated than their wild ancestors. It is not uncommon to hear that sheep, cows, and dogs are dumbed-down versions of wild animals. When I introduced Darwin and Wallace in chapter 3, I emphasized how strikingly similar their writings on evolution and natural selection were. They both emphasized the struggle for existence, in which only individuals with the most favorable traits survive, reproduce, and pass on those traits to their offspring.

What I neglected to point out, though, is that Darwin and Wallace had wildly different views of domestication. In their 1858 writings, both refer to the domestication process, by which humans create specific breeds of dog, sheep, and plant that suit their needs and preferences. In Darwin's view, domestication provides the perfect showcase of selection. He follows this point in the *Origin of Species*,[1] in which he devotes the entire first chapter to domestication. He does so intentionally. No one at the time could deny that humans had been able to make extreme changes to wild species through the process of artificial selection. Darwin argues that if humans can create different breeds of pigeons or dogs from a single common ancestor, then nature can do so much better as it is "immeasurably superior to man's feeble efforts"—and has had a lot more time to do it.

Wallace agreed that domestication provides proof for selection, but he emphasized that the traits selected by people provide no benefit in nature. He saw domestication as a weakening exercise. He argued that the struggle for existence continuously tests animals, that it keeps their muscles strong and their senses heightened. By providing food, shelter, and protection from predators, humans—through domestication—create a bunch of weaklings that would stand no chance in nature. "Half of its senses and faculties are quite useless," he says of the domestic animal. "Our quickly fattening pigs, short-legged sheep, pouter pigeons, and poodle dogs could never have come into existence in a state of nature, because the very first step towards such inferior forms would have led to rapid extinction of the race."[2]

The difference between Darwin's view and Wallace's view on domestication is, of course, a difference in focus. Where Darwin assessed the process of selection and highlighted the parallels between artificial and natural selection, Wallace argued that any traits that humans select in their domesticated animals would be quite useless in nature. Interestingly, however, this Wallace-like thinking, that domestication produces animals that are useless, or even downright dumb, has pervaded much of Western science. And that thinking has blinded scientists, and resulted in agricultural practices in which animals are not able to medicate themselves. This has led to more disease, and the need to treat animals with vast amounts of drugs. As a result, we are now experiencing widespread antibiotic resistance that is plaguing animals and humans alike. But not all is lost. As I describe in this chapter, domesticated animals have *not* lost their ability to medicate themselves. And by knowing this, we can provide these animals with the means to improve their own health.

GUT FEELING

To better understand the role of animal medication in agriculture, I'd like to introduce you to Fred Provenza, professor emeritus of behavioral ecology in the Department of Wildland Resources at Utah State University. Like so many other scientists, he used to believe that domesticated animals can no longer look after themselves and have lost their ability to regulate their own diets. But his views changed dramatically during his PhD studies conducted in Cactus Flat, in southwestern Utah. On the Fourth of July, 1976, while friends and family were celebrating the two-hundredth anniversary of America's independence with hamburgers and hot dogs, he and his wife, Sue Provenza, drove down from Utah State University to the desolate landscape of Cactus Flat. His goal: to study whether domesticated goats can stimulate the growth of lush vegetation in the hot and dry landscape to make it suitable for grazing by mule deer, bighorn sheep, and domesticated cattle. So, the Provenzas built fences to create six pastures; in each one of them they released fifteen Angora goats, which they leased from the Navajo Nation.

The landscape in Cactus Flat is dominated by blackbrush (*Coleogyne ramosissima*), a medium-sized shrub in the rose family, the bark of which turns black when wet, hence the name. It is not an enticing plant, with most of the shrub consisting of dry, tough, and woody branches. It is only the young outgrowths at the end of the branches that are green and lush.

"I would look at that plant and think, how can any creature survive on it," says Provenza.[3]

Researchers at Utah State had observed that there were more nutrients in the young shoots: they had more energy, protein, and minerals than the old woody parts.

"And so they were thinking, could we use goats to feed on blackbrush to stimulate these new twigs and make it better country for wildlife species and cattle?"

Because the young shoots are way more nutritious than the woody parts of the plant, the Utah researchers, Provenza included, predicted that the goats would prefer those young shoots and avoid the woody stuff. He remembers his excitement when finding a bunch of shrubs with tons of green shoots in one of the pastures. Provenza herded his goats toward them, expecting them to feast, bleat with excitement, and maybe even show him some gratitude. But the goats just stood there, staring at him as only goats can do. They eventually turned their backs and started grazing the woody branches of nearby shrubs. Provenza was perplexed. It was not at all what he expected. As he analyzed his data, the results were clear, though: goats avoided the young growth, and made their living eating the woody and tough parts of the brushes instead.

"Watching the goats in southern Utah was so intriguing," he says. "They must know something that we don't!"

Provenza recalls relating his observations to a professor of toxicology. The reply was: "that just goes to show domesticated animals lack nutritional wisdom." In other words, *there was no way* that goats made any sensible decisions when it came to their diet.

But it turned out the goats were smarter than that toxicology professor. They knew that the quality of their food depends not only on the nutritious substances they contain but also on their secondary compounds. As we saw previously, plant scientists distinguish between primary and secondary plant chemicals.

Primary chemicals are those that are needed for nutrition, including carbon sources for energy, nitrogen sources for protein, and minerals and vitamins needed for sustenance, growth, and reproduction. Secondary chemicals are all other chemicals. As I described before, not understanding what these were for, many scientists assumed they were simply waste products and served no function. However, just when Provenza was hammering fence posts into the hot dirt in Cactus Flat, ecologists started realizing secondary compounds are essential for plant life. As we saw earlier, plants use them to deter herbivores and parasites, attract pollinators, provide colors to flowers, protect against dehydration and sunburn, and kill competitor plants. With that in mind, Provenza hypothesized that some compounds in the fresh blackbrush must be deterring the goats.

Provenza started extracting the secondary compounds of blackbrush shoots and mixing them with food pellets. He fed the pellets to experimental goats, one compound at a time. He failed to find any difference until finally testing the very last compound: condensed tannins, a class of chemicals that cause lesions in the animals' gut mucosa and inhibit digestion. Tannins occur in tea and red wine, where they are responsible for the astringent feeling when drinking them—this is based on cells and tissues contracting. These chemicals make up 70 percent of the bark of young shoots. And indeed, the tannins turned out to be responsible for the goats' aversion. Interestingly, the goats happily ate the pellets with tannins at first, but started avoiding them the next day.

What this taught Provenza was that goats avoid tannins. But not innately. Instead, they need to experience the negative effects of tannins before avoiding them. They need to learn. In goats, tannins cause nausea. Blackbrush plants load their young and nutritious shoots with tannins, so herbivores avoid them.

By eating these shoots, goats feel sick, and this teaches them to avoid eating those shoots in the future (a win for the plant!). If this was really the case, Provenza reasoned, then giving the goats something that relieved them from nausea should stop their aversion of tannins. So he went back to Cactus Flat and fed the goats polyethylene glycol, a polymer derived from petroleum that is better known as PEG. PEG is a harmless compound that is commonly used as an ingredient in medications. In the digestive tract of goats, PEG binds with the tannins, so the goats no longer experience nausea. As Provenza expected, PEG-fed goats started happily munching the young shoots of blackbrush.

These studies showed that goats can intentionally avoid consuming harmful chemicals, and that they do so by learning to associate eating particular foods with feeling sick. In other studies, Provenza found that energy-deficient lambs preferentially eat foods that they associate with starch,[4] and that lambs deficient in specific minerals learn to prefer foods with those minerals. Provenza also exposed sheep to nauseating lithium chloride during anesthesia. Despite being unconscious when experiencing the nausea caused by the chemical, the sheep subsequently avoided the food they had most recently eaten.

In another study, Provenza studied acidosis, a common ailment in sheep, goats, and cattle. Acidosis happens when livestock eat too much grain. The overload of grain results in a big influx of carbohydrates. Bacteria in the rumen ferment the carbohydrates and produce lactic acid, which slows down the gut and can result in dehydration and even death. Acidosis also occurs in people, and is often treated with sodium bicarbonate. Therefore, Provenza fed two groups of sheep large amounts of barley to initiate acidosis, then gave them choices between either onion- or oregano-flavored rabbit pellets with or without sodium bicarbonate. Impressively, the sheep learned to associate the particular pellets that coincided with relief from acidosis, and started choosing those pellets in choice tests.[5]

Thus, goats and sheep make both positive and negative associations between foods and gut feelings, resulting in preferences and avoidance. Most impressively, it is their bodies and their ancient limbic brain that manages emotions—not the cognitive cortex of their brains—that tell them what to do. In this sense, the animals listen to their gut. Despite being pampered by humans for some ten thousand years, domesticated animals apparently have not lost their ability to listen to their bodies and regulate their own diet. They are not as dumb as Wallace and that toxicology professor may have expected.

Provenza calls it nutritional wisdom. "It is the wisdom at the level of cells and organ systems to know what they need to nourish themselves," he says. "Not only primary compounds, but this vast array of secondary compounds."

In his fascinating and acclaimed book *Nourishment*,[6] Provenza describes how his four decades of research have demonstrated that the animal body can make some rather sophisticated choices when it comes to nutrition. And those sophisticated choices include self-medication.

WORMS AND TANNINS

Livestock face many challenges, but parasitism is one of the most urgent. Worms that infect the gut often cause poor growth and death. But the drugs that are used in livestock are failing, as pathogens and parasites are evolving resistance, which is bad for people and animals alike. So, a logical question for Provenza was whether livestock can medicate themselves when infected with worms. If livestock could medicate themselves, it might mean that we could stop overusing human-made antiparasitic drugs and hopefully slow down the parasites' ability to evolve drug resistance. He would address this question when then-graduate student Juan Villalba joined his lab in 1993.

Villalba was trained in biochemistry and obtained his master's degree studying ruminant nutrition at the Universidad Nacional del Sur in Bahía Blanca, Argentina.

"Health was not really that interesting to me at the time," says Villalba.[7]

But when Provenza showed that sheep can treat their own acidosis, Villalba was hooked. This was also the time when worms evolved high levels of resistance to anti-worm drugs (such as ivermectin), and the need arose to explore other ways to fight off parasite infections. Joining Provenza's lab turned out to be a master move, with the two scientists starting a decades-long collaboration to study how livestock can medicate themselves and what we can learn from them to improve agriculture.

In 2006, Villalba, Provenza, and their colleagues showed for the first time that sheep can use medication against parasites.[8] Ironically, it turned out that tannins, those chemicals that nauseated the goats in Cactus Flat back in 1976, could act as an effective worm medicine. Studies had shown that livestock feeding on tannin-rich plants, such as sulla (*Hedysarum coronarium*) and sericea lespedeza (*Lespedeza cuneata*), experienced fewer worms and better growth. This happened because tannins are toxic to worms and because they boost intestinal immunity. Villalba and Provenza knew of these studies. They had also learned about Michael A. Huffman's work on chimpanzees.

"Tannins are a bit like the *Vernonia* plants that chimpanzees use to medicate themselves against parasites," says Villalba. "Tannins do not kill sheep, but they cause lots of negative effects. It is the same with *Vernonia*: they are very toxic plants that can actually kill chimpanzees."

Thus, both tannins and *Vernonia* can be seen as medicines with harmful side effects: useful when infected but to be avoided when not needed.

Villalba and Provenza set up an experiment in which they compared sheep that were uninfected with sheep that were infected with worms. They offered the sheep grape-containing pomace supplements with condensed tannins and tracked tannin consumption and worm burdens over time. Over the first twelve days of the experiment, the infected sheep ate significantly more of the tannin supplement than the uninfected sheep. As a result, their worm infections subsided. At that point, the sheep started avoiding the tannins again, just as the researchers had predicted.

In a follow-up study,[9] the research team cleared 22 lambs of parasitic worms with a mix of anthelminthics (anti-worm drugs), then infected them with a standard dose of *Haemonchus contortus*, a common parasite of livestock known as barber's pole worm. Sheep received either beet pulp supplemented with tannins, or beet pulp without tannins. Later, the sheep were given choices between the two types of beet pulp. As predicted, the sheep that had been fed tannins during worm infection preferred the tannin-containing beet pulp, while the other sheep avoided it. When parasitic worms were cleared with anthelminthics, the sheep again avoided the tannin-containing pulp. Altogether, these results suggest that worm-infected sheep associate tannins with positive effects (reductions of fecal egg counts were observed), and that they increase their tannin intake to reduce worm burdens—despite the nauseating taste. When that is no longer needed, they stop eating the tannins,

FIGURE 9.1. Juan Villalba tends to sheep, offering them multiple foods that they can use as medication; the sheep choose their own diet. Photo by Ravell Call, Deseret News.

which makes sense given the nausea they experience when consuming them. All in all, sheep seem to make some rather sophisticated diet choices.

Today, the use of medicinal plants by livestock is well supported. One group in Israel found that worm-infected Mamber goats consumed more foliage of the tannin-rich shrub *Pistacia lentiscus* than uninfected goats when given the choice between that plant and hay.[10] Researchers from France and Mexico discovered that sheep infected with *H. contortus* ate more food containing tannin-rich *Lysiloma latisiliquum* plants than uninfected sheep.[11] A study in Scotland showed that compared to uninfected lambs, lambs infected with the worm *Trichostrongylus colubriformis* consumed more high-protein food when given a choice between different diets. The researchers hypothesized that the lambs did this to compensate for the worm-induced stunting of growth and the high costs of immune responses against the infection.[12] And in another experiment,[13] Villalba found that sheep infected with *Haemonchus contortus* increased their consumption of a diet supplement that contained polyphenols and terpenes from grape, olive, and pomegranate. These chemicals are antioxidants and most likely act to reduce inflammation. As a result, the sheep managed to grow at higher rates despite suffering from worm infections.

What is even more impressive is that livestock can learn to use specific medicines for specific ailments.

"People take aspirin to relieve headaches, antacids for stomach aches, and ibuprofen for pain," says Provenza.

But until recently, no one had ever demonstrated that animals can be similar all-around doctors. So, Provenza and Villalba conditioned sheep to use three medicines that diminish illnesses: sodium bicarbonate to alleviate acidosis from eating too much grain; PEG to counteract the adverse effects of tannins; and dicalcium phosphate to counteract the adverse effects of eating food with high levels of oxalic acid. They then offered sheep either grain or food with tannins or food with oxalic acid and gave them access to the three medicines. Sheep chose the medicine that rectified the specific malady, thus showing that herbivores can learn to treat specific illnesses by selecting foods with appropriate chemicals.

FOODSCAPES

As the research by Provenza, Villalba, and others has demonstrated, livestock show an astonishingly refined palate. They learn from past experiences to meet their needs for nutrients and medicines, as they nibble their way through the day. However, many of our agricultural systems are still designed on the premise that domesticated animals are dumb, that they cannot regulate their own diet, and that we, humans, are superior at providing them with the foods they need. That's why the livestock industry likes to use total mixed rations: food mixes that are designed to provide a mix of all the nutrients that an animal needs to obtain the fastest growth at the lowest cost. These mixes generally contain primary chemicals but no secondary chemicals. To take care of parasites and pathogens, drugs are simply added to the diet.

"The animal has no choice whatsoever," says Provenza. "Those rations are an affront to animal rights and freedoms. I often ask people: how would you like it if someone formulated a ration for you and that's all you had to eat, every day, all day long? How long could you stand that?"

A major problem, says Provenza, is that the rations are prepared for an *average* animal, and the animal has no choice whatsoever in composing its own diet. But no animal is average. Every animal is unique, and each animal, when given the choice, will assemble its own diet given its individual and unique needs. That is what Provenza found in a study of beef calves. He fed sixteen calves a total mixed ration of rolled barley, rolled corn, corn silage, and alfalfa hay. Another fifteen calves were given the same four foods in separate troughs so the animals could assemble their own diet. Think of it this way: the mixed ration calves were given an á la carte menu with only one dish to choose from, while the choice animals had access to a whole buffet.

Over the course of two months, each of those fifteen buffet calves assembled its own unique diet, each one having different relative amounts of the four foods. That is what Provenza expected. What he did not expect is that, overall, the buffet calves ate less food than the calves fed the total mixed ration. Yet, they gained just as much weight (beef!). Provenza estimated that the cost of producing a kilo of beef went down by 20 percent when calves could assemble their own diet.[14]

"That was amazing to me," he says. "It showed if they have choice, they can better meet their needs. They do not need to over-ingest food to obtain the nutrients they are lacking."

(There are parallels here with humans. Provenza thinks that the current obesity epidemic in Western civilization is to a large extent because humans are overeating low-quality processed food dominated by carbohydrates to obtain enough of the limited protein sources and minerals in this food.[15])

Ironically, with the industrialization of food production, we have progressively failed to enable livestock to self-medicate. Secondary chemicals, rather than being seen as the natural pharmacy they are, are viewed as poisons, to be avoided at all costs. Livestock are fed standardized food rations, consisting of mainly carbon and nitrogen sources, with no chance to self-medicate if needed. Instead, we treat the animals with various medications, including anthelmintics and antibiotics. Our livestock can medicate themselves. We just don't let them.

That's why Provenza and Villalba believe that it is essential to rethink the livestock industry, and once again take advantage of the nutritional wisdom that our domesticated sheep, goats, and cattle have retained. To do that, we need to, first and foremost, provide choices so animals can assemble their own optimal diets and medicate when needed, without the need for large-scale application of antibiotics and anthelminthics—whose overuse has resulted in widespread resistance. That requires giving them the tools to learn from trial and error. Moreover, animals need to live in social groups. In one study, Villalba and Provenza compared the preferences for consumption of PEG by lambs that were fed tannin-rich diets and were raised alone or raised with their mother. They found that lambs develop stronger preferences for PEG, and eat more of it when raised together with their mothers than when raised alone.[16]

"If mother is consuming medicines," says Villalba, "the animals will be more likely to use the medicines. The mother is really important."

This is not because the lambs copy their mothers' behaviors, but because both mothers and lambs are more prone to trying out novel foods in the presence of each other. In other words: when living together, they have less neophobia than when living alone, as discussed in chapter 8.

Sheep, goats, and cattle are social animals, and in their social groups, they can also learn from each other. Back in that year 1976, Provenza learned this firsthand. The Cactus Flat landscape, dominated by blackbrush, also sprouts juniper trees. At the base of some junipers, wood rats build their houses (in which they incorporate aromatic plants for fumigation, as we saw in chapter 5). The houses look a little like beaver dams—minus the surrounding water.

"The goats in one pasture started eating woodrat houses," says Provenza. "Why on earth are they doing this?"

Woodrat houses have different rooms. One is the bathroom, and it is soaked in urine. Apparently, these goats discovered the bathrooms to be great sources of nitrogen. When Provenza weighed the goats at the end of the season, they had not lost nearly as much weight as the goats in the other five pastures. Only one group of goats figured this out, and Provenza thinks it is because one goat discovered it, and the rest copied it. The lesson, Provenza thinks, is that it is not just important to allow offspring to learn from their mothers but to maintain livestock in social groups where newly discovered medication behaviors can be readily shared.

Giving choice, letting mothers spend time with their offspring, and allowing livestock to live in social groups—those are the ingredients needed to reintroduce self-medication in our livestock. Villalba thinks that one way to do that is to let animals roam outside as much as possible.

"The main thrust of my research is to have fewer animals in barns and more in pastures," he says. "We can improve the quality of our pasture lands and range lands. Feed the animals in a much more natural system without the need of grains or confinement."

To that end, Villalba's current research focuses on designing diverse pastures that are not just good for the animals but also for the environment. Studies are now piling up to show that mixed diets benefit livestock by allowing them to obtain balanced diets, by letting them medicate themselves when needed, and by reducing stress (animals get stressed when eating the same food all the time).[17] Cows raised in pastures with different plant species produce more milk than those raised in less diverse pastures.[18] And environmentally, mixed diets reduce the production of the greenhouse gases methane and nitrous oxide, while redirecting nitrogen into feces and away from urine. This makes for better fertilizer and less nitrogen runoff.[19] Moreover, more diverse ecosystems are more resilient to climate change, so creating diverse pastures also buffers farmers against the negative effects of droughts and flooding.

Changes to the livestock industry may seem daunting. But farmers could start small and improve their systems one step at a time. To that end, Villalba is creating diverse paddocks with a mix of plants that provide food, medicine, and antioxidants.

FIGURE 9.2. An experimental field divided into sections with different plants. Experiments by Fred Provenza, Juan Villalba, and their colleagues have shown that sheep and other livestock can learn to use medicinal foods, as long as they have access to a diversity of plants. Photo by Ravell Call, Deseret News.

The idea is not even to change whole fields but to create islands of diversity within existing pastures. The other good news is this: while free-ranging ruminants may eat up to fifty different plant species a day, most of their diet comes from three to six species. And providing that number of species can go a long way to improving livestock health.

"I like to call these pastures foodscapes as opposed to landscapes," he says. The animals can use medicinal plants when they are infected. "But we don't have to wait till the animals get sick. Even with small doses, antioxidants make a huge difference: the animals have better immunity, and suffer less from bacteria and parasites."

Thus, diverse landscapes provide animals with both prophylaxis and therapy.

While these changes may be relatively straightforward to realize in ranches and pastures, many animals are maintained in pens, sheds, and barns. But even in such confinement, Villalba thinks that the tools are available to provide livestock with crucial food choices.

"The livestock industry is already developing targeted treatments," he says. "Treat the animal that needs it. Not the massive deworming that they do with thousands of sheep or thousands of cows."

In modern operations, animals step on a scale, and a tag shows the change in weight. If the weight gain has been fine, one chute with food opens. If not, another chute opens. Systems like this can be adapted to provide livestock with different types of food and to move away from the single total mixed ration loaded with antibacterial and anti-worm drugs.

Provenza and Villalba think that, at the end of the day, transforming the livestock industry will not only be good for the animals and reduce our dependence on failing drugs, but will also generate more income and be better for the climate.

"A more diverse landscape results in less methane production," says Villalba. "But what matters for the producer? Your calves will be bigger. Your animals will be healthier."

Giving animals choices will also improve the quality of our food. The plants that herbivores eat are loaded with chemicals that bolster their health. The chemicals protect animals against diseases and pathogens through antimicrobial, antiparasitic, immune-boosting, and anti-inflammatory effects. In turn, human health is linked with the diets of livestock through these chemicals. The benefits to people accrue as herbivores assimilate some plant chemicals and convert others into metabolites that become part of muscle, fat, and milk.

"This rich tapestry of compounds promotes human health," says Provenza. "This is very obvious to Indigenous cultures. We feed what feeds us."

Thus, by taking care of our livestock, we take care of ourselves.

And it is not just livestock and humans whose lives we can improve. As we will see in the next chapter, we can equally apply the principles of animal medication to help our suffering honey bees.

10

Sticky Bee Business

A picture says more than a thousand words. That's why, in 2013, the American supermarket Whole Foods teamed up with the Xerces Society, a nonprofit organization that protects wildlife through the conservation of invertebrates and their habitat. Their goal: to showcase a supermarket in a world without bees. So, in the produce department at the Whole Foods Market at University Heights in Rhode Island, employees removed all those products that depend on bees. Without bees, we would lose apples, onions, avocados, carrots, mangos, lemons, cantaloupe, cucumbers, kale, leeks, and many, many other items. Before and after pictures show a shocking difference, with half of the shelves empty and desolate. What these photos make clear is that bees are needed for so much more than honey. Indeed, scientists estimate that bees and other pollinators are responsible for one in three bites of all the food we consume. This food even includes milk and yogurt, since the main food for dairy cattle is bee-pollinated alfalfa and clover.

Honey bees are the most important managed pollinators worldwide. They contribute to the production of thirty-nine of the fifty-seven leading crops grown for human consumption, accounting for hundreds of billions of dollars in agricultural

FIGURE 10.1. Whole Foods produce section in a world with bees. Photo by Whole Foods.

FIGURE 10.2. Whole Foods produce section in a world without bees. Photo by Whole Foods.

value.[1] Many of these crops are grown in large monocultures, and require pollination during one short period each year. While many wild bees, beetles, flies, butterflies, birds, and bats pollinate flowers, honey bees are the only suitable pollinator for large-scale pollination. This is because they can be bred in huge numbers, maintained throughout the year, and transported around regions to pollinate where they are needed.

In the United States, many beekeeping businesses are migratory, with beekeepers loading thousands of colonies on semitrucks and driving them around the country from crop to crop. In a typical year, a migratory beekeeper may truck their bees to pollinate melon vines in Florida, apple trees in Pennsylvania, blueberry bushes in Maine, and almond trees in California. Most of us are blind to this practice, and happily buy melons or almonds without realizing the hard life of beekeepers and their bees. That is—until the unlucky moment that one of those semitrucks crashes on a highway, releasing millions of angry bees and making headlines in local newspapers.

The fruits, nuts, seeds, and vegetables that honey bees pollinate not only account for one-third of the amount of food we consume but also provide the bulk of the micronutrients that we need. Thus, losing bees means losing human health. Sadly, bees are in trouble—and by extension, so are we. In recent decades, beekeepers in the United States and Europe have suffered annual losses of 10 percent to 50 percent of bee colonies,[2] as a result of a lack of flowers, pesticides, parasites, and pathogens.[3] Overall, the number of total hives in the United States has almost halved since the 1940s, from close to 6 million to around 3 million.[4] Like humans, bees are social animals who live huddled together in cities and metropolises at densities that are ideal for the spread of infectious diseases.

As the COVID-19 pandemic told us, the price of sociality can be high.

But social interactions can also provide opportunities for medication. As we will see in this chapter, honey bees are masters at social medication. And if beekeepers allow their bees to carry out their highly evolved behaviors, we can go a long way in saving bees—and human health at the same time.

CRIPPLED WINGS

Bees suffer from a freaky collection of parasites and pathogens that are beyond the imagination of even the most creative artists of the Star Wars franchise. Take *Varroa destructor*, an eight-legged mite that latches onto a bee, inserts its sucking mouthparts through the skin into the fat body (sort of the bee's equivalent to our liver), and then happily gorges for days on end. To put its size in perspective: *Varroa destructor* would be as big as a donut-sized tick-like creature stuck to a human belly. Now consider the fact that bees may carry two or three of these mites at the same time. Beyond sucking the life out of the bee, these mites suppress the bee's immune system, and inject deformed wing virus into its blood. As the name implies, infection with this virus can result in crippled wings, preventing the poor bee from flying.

Other viruses that infect bees have telling names such as acute bee paralysis virus and chronic bee paralysis virus. Bees also suffer from fungi, such as the deadly chalkbrood-causing *Ascosphaera apis*, which makes bee larvae look like overripe pieces of French Brie. And then there is *Nosema ceranae*, a highly contagious single-celled parasite that decimates adult bees. Bees also suffer from tracheal mites (which inhabit the trachea, the bee's lung-like organ) and a slew of bacterial pathogens that liquefy bee larvae.

While all these creatures are detrimental to honey bees on their own, their combined pressure is especially devastating. Together with the stresses of agricultural pesticides, and the malnutrition caused by agricultural monocultures (like livestock, bees benefit from a varied diet and prefer ecosystems with many different plants), these parasites and pathogens are responsible for the recent declines in honey bees.

Marla Spivak, Distinguished McKnight University Professor of entomology at the University of Minnesota, summed it up during her TEDGlobal talk in 2013: "I don't know how a bee feels with blood-sucking parasites and with viruses. But I know what I feel like when I have the flu. What if I lived in a food desert? And then I consumed food that contained enough of a neurotoxin that I could not find my way home?"

Disease pressures have significantly increased since humans began domesticating bees because of the much higher densities at which bees are maintained.[5] Wild bees typically live in colonies of up to eighteen thousand bees in hollow trees, and at a density of one such colony per square kilometer. In contrast, modern commercial beekeeping operations typically maintain bee colonies of twenty thousand to fifty-two thousand bees in rectangular boxes with multiple frames on which bees can build their wax comb. Industrial operations may maintain apiaries with thousands of such boxes, placed together in small areas.

But even at natural densities, honey bees are highly social animals, prone to infection. As a consequence, bees have evolved sophisticated defense mechanisms to protect themselves against the increased risks of social life. Like other insects, bees have a number of immune responses, which include the production of antimicrobial peptides that kill bacteria.

Surprisingly, however, the immunological arsenal of bees is smaller than that of other insects, such as malaria mosquitoes and fruit flies.[6] The major hypothesis for this lack of immune genes is that bees protect themselves through other defenses instead.

Bees engage in what scientists like to call social immunity—behaviors by which individuals work together to combat disease.[7] Some bees act as guards, protecting the hive entrance and barring diseased bees from entering and spreading disease. Other bees patrol the hive and remove bee cadavers, to prevent decay and spread of pathogens. This is the hygienic behavior, controlled by a few genes, that we saw in a previous chapter. Bees can also work together to raise the temperature of the hive and kill disease-causing bacteria: they do so by vibrating their flight muscles to generate heat.[8] And then, of course, honey bees are experts in medication. By lining their colonies with medicinal tree resins, honey bees keep pests at bay, enhance their microbiome, and increase their survival and honey production.

THE BANE OF THE BEEKEEPER'S EXISTENCE

If you have ever opened a bee box, you will know that honey bees add sticky stuff to their hives. This sticky stuff is called propolis, and bees make it by mixing resins with wax. As you may remember from a previous chapter, resins are the secretions that some plants use to protect themselves against pathogens and insects and that help them recover from wounds. When tree resin fossilizes, it can turn into amber. If you are old enough, you may remember the vivid scene in the 1993 movie *Jurassic*

Park. A mosquito full of dinosaur blood is captured in the resin dripping from a tree; and thereby unwittingly becomes the source of the DNA from which businessman John Hammond and his scientific team recreate extinct dinosaurs.

Honey bee societies are highly organized, with different bees playing specific roles. The queen bee lays the eggs. Eggs develop into larvae, then into pupae, and finally into adult sons or daughters. Sons are known as drones. Their only role in life is to mate with other queen bees. (They don't do any work, and they also do not sting, which is why I always send angry looks to fellow parents who, when they see a bee or a wasp, warn their kids that "he" may sting). Daughters are the worker bees: they are the ones that keep the colony afloat and make sure all tasks get completed.

Different worker bees specialize on different tasks. Just as there are bees that feed larvae (nurse bees), and bees that collect pollen (the bees' source of protein) or nectar (to produce honey), some workers are specialized in collecting resin. These bees collect resin droplets from the bark, fruits, buds, and young leaves of plants. This involves an intricate behavior that lasts about seven minutes: the bee uses her mandibles and forelegs to break off the resin, transfers it to her middle legs and then moves it into the pollen baskets on her hind legs (as the name implies, the pollen basket is a structure on the hind legs that other specialized bees use to collect pollen).

On her return to the hive, the bee cannot remove the resin herself, as it is too sticky. Instead, other worker bees, called cementing bees, bite off little pieces of resin, smooth it with their mandibles, and add wax. At this point, the mix of resin and wax is called propolis. In temperate regions, bees mostly collect resin from *Populus* trees (poplars, aspens, and cottonwood), but they can also use pines, birches, alders, beeches, and horse chestnuts.

FIGURE 10.3. A honey bee returns to her colony with a fresh supply of resin on her hind legs. Honey bees use resin as medication. Photo by Eric Tourneret.

In tropical regions, bees prefer other trees (including *Clusia minor* and *Clusia rosea*) and shrubs, including *Baccharis dracunculifolia*, a plant well-known for its medicinal properties.[9]

One of the leading experts on bee medication is Marla Spivak, the University of Minnesota professor who we just met through her TED talk on the plight of honey bees. In addition to her distinguished professorship, Spivak is a MacArthur Fellowship recipient for her groundbreaking work on honey bee health. Spivak tells me that those honors were unexpected, as her lifelong career in honey bee health only started by accident in 1973.[10] She was a first-year undergraduate student and, like most undergraduate students, "bored and without direction." As this was pre-internet and pre–social media, she decided to go to the library for something to read, and "picked up something random." The random book was *Bees' Ways,* a natural history book by George DeClyver Curtis. Spivak started reading and got hooked.

The school Spivak attended, Prescott College in Arizona, was an experiential school, with lots of outdoor environmental education. Having told her advisor she was interested in bees, he hooked her up with a commercial beekeeper who was also an organic farmer. After learning what she could at the farm and then in a research position at the US Department of Agriculture (USDA), she went on to earn a PhD from the University of Kansas. In 1993, she went back to the USDA lab as a postdoctoral researcher to study hygienic behavior, by which worker bees remove dead and diseased larvae and pupae from the colony. She first became interested in propolis when she was a professor at the University of Minnesota and was visiting with colleagues in Brazil.

There, she noticed that people go to Walmart to buy propolis tinctures, which they use for anything from a sore throat to a cold or a mouth infection. At the same time, researchers at the University of Minnesota's medical school were looking at alternative treatments for HIV infection in humans. One colleague of Spivak's, Genya Gekker, obtained some propolis for a sore throat from a local farmers' market. Arriving home, however, Gekker decided to leave her throat as it was, and she used the propolis for her research instead. Adding propolis extracts to human brain cells that Gekker maintained in flasks in the lab, she discovered that propolis stopped HIV from entering the cells.[11] (This is similar to the antibodies elicited by COVID-19 vaccines, which stop SARS-CoV-2 virus from entering cells in the human airways.) Gekker and Spivak authored a paper on this work, and Spivak quickly discovered that propolis had been used by humans as medicine for thousands of years.[12] In addition to the antiviral activity shown by Gekker, propolis has antitumor, antioxidative, antimicrobial, and anti-inflammatory properties.

Yet, no one seemed to think that bees may use propolis as medicine. In the classic bee book *The World of Bees*, author Murray Hoyt describes the role of propolis as a building material.[13] Propolis is a glue-like substance that bees use to stick things together, smooth surfaces, and fill up holes and crevices: a handy material for bees that inhabit the rugged insides of hollow trees. They also use it to encapsulate animals that invade the bee colony but are too big to be carried out: hives even harbor mummified mice covered by propolis. Propolis is a Greek word, and means "in front or defense of the city." This name comes from the fact that bees use propolis to shape their nest entrance into a small gate that allows entry of bees but keeps enemies out.[14]

Because propolis is sticky, beekeepers do not like it. While Hoyt describes the use of propolis as essential to honey bees' lives in hollow trees, he states that "propolis is not necessary to apiary life; in fact it is the bane of a beekeeper's existence."[15] Beekeepers regularly remove propolis from hives during routine control. In beekeeping classes, instructors will often start by showing how to remove propolis. And by picking out colonies that produce less propolis for their operations, beekeepers have also selected against propolis over time. Paradoxically, we now know that bees use propolis as an essential defense against pathogens. By getting rid of this medicinal substance, beekeepers have unwittingly removed a major component of disease control.

MEDICINES AND MICROBES

Enter Mike Simone-Finstrom, a graduate student who joined Spivak's lab just as she became interested in propolis. Simone-Finstrom grew up near Chesapeake Bay, where he was drawn to the water, and carried out many a science fair project on fish. He was convinced that one day he would become a marine biologist. To realize this dream, he entered graduate school at the University of Minnesota and joined a fish lab. But things did not work out quite as expected. Taking a course on insect behavior, he realized that insects, especially those living in big social groups, are at least as cool as fish. He quickly fell in love with bees and, after a chat with Spivak, decided to switch his research from fish to bees.

Simone-Finstrom's first experiment in Spivak's lab was to test the effects of propolis on the growth of bacteria that cause American foulbrood. This disease is one of the most contagious bee diseases there is and is caused by the bacterium *Paenibacillus larvae*. Nurse bees can spread the bacteria to larvae as they are feeding them. A mere ten bacterial spores are enough to cause an infection in young larvae, and rapid bacterial growth in the gut results in billions of bacterial offspring. The larva then turns into a semiliquid mass. To detect the disease, beekeepers push a matchstick in the brood cell and slowly withdraw it. If this results in a three-to-five-centimeter, dark brown ropy thread, it is American foulbrood. (Is this gruesome enough for you? If not, please note that if the thread is shorter and light grey, it is more likely to be European foulbrood, another bacterial disease.[16])

Together with a colleague from the Ezequiel Dias Foundation in Minas Gerais, Brazil, Simone-Finstrom extracted chemicals from propolis and added them to agar plates seeded with

Paenibacillus larvae bacteria. Many of the extracts showed clear zones in which the bacteria could not grow, demonstrating that the extracts slowed bacterial growth or killed them.[17] For Simone-Finstrom, this was supposed to be a side project. But these exciting results, and his attendance at an International Union for the Study of Social Insects meeting in Washington, DC, made him decide to stick with the sticky propolis.

Simone-Finstrom was at the meeting with Spivak, and they heard a presentation by Michel Chapuisat, then a junior investigator at the University of Lausanne in Switzerland. In it, Chapuisat presented the research he had done with Philippe Christe on wood ants. As we saw in chapter 6, these insects use tree resin as a form of medical prophylaxis: incorporating little pieces of resin into their nests reduces the presence of disease-causing microbes.

At this point, Spivak turns to Simone-Finstrom and whispers excitedly: "Mike, this is it!"

The Swiss research on ants opened up a whole new avenue, with bees potentially using propolis as medication. Until that point, all research on propolis focused on the health benefits for humans. Everyone simply ignored the health benefits for bees.

"We thought we knew what it was for," says Spivak. "It was for cement, moisture control. We just stopped there."

Spurred on by the Swiss research on ants, Spivak and Simone-Finstrom embarked on a series of studies to test whether bees can use propolis as medicine. In one study, they painted propolis on the inside walls of bee hives and measured the amount of bacteria as well as immune function of bees.[18] Compared to control colonies in which they did not apply propolis, the added propolis led to reduced antibacterial immune responses in bees. They hypothesized that the reduced immune response was due to the reduced bacterial loads in the colony. Immune responses are costly, and adding propolis allowed the bees to save the energy they would otherwise invest in these defenses.

In another study, Simone-Finstrom and Spivak set up bee colonies and either added spores of the chalkbrood-causing fungus *Ascosphaera apis*, or left colonies uninfected.[19] They then monitored the entrances to the hive boxes to count the number of resin-collecting bees, and noticed that infected colonies sent out more bees to collect resin.

"It was a nice result," says Simone-Finstrom.[20] "But I thought it was a fluke. I simply did not believe bees could actually do this. So we did it a second year, and saw the same thing. And then we did it yet another year."

Over the course of three years, Simone-Finstrom and Spivak saw that the number of resin-collecting bees went up by almost 50 percent.

Another experiment in which they compared colonies with added propolis to colonies without propolis showed that resin-rich colonies experienced less infection after challenge with the chalkbrood fungus. Chalkbrood is a larval disease and does not affect adults. Simone-Finstrom and Spivak's results thus showed that adult bees respond to chalkbrood infection in larvae by collecting more medicinal resin and incorporating it into their nests. A fine example of social rather than self-medication.

"I think it is just a cool thing," says Simone-Finstrom. "You have less than 1 percent of all foragers collecting resin. And they are responding to a larval infection. It is not like they are sick themselves. How is that happening? Is it the odor of the chalkbrood? We don't know yet."

Studies in Italy and Germany have shown that honey bees may also use propolis as a defense against *Varroa* mites and the viruses they transmit.[21] Bees even line the brood cells where pupae develop with propolis, which kills some of the mites and reduces their fertility.[22] In another study, a graduate student in Spivak's lab set up experimental colonies with and without propolis and then seeded them with the pathogenic bacterium *Paenibacillus larvae*. Impressively, bees in propolis-rich colonies produced larval food with antibiotic compounds, and these colonies experienced less American foulbrood disease.[23] Whether the bees produced these compounds, or whether the compounds leached into the honey from the propolis itself, remains an open question.

Propolis does not only work as a medicine against parasites and pathogens, but also helps bees develop a healthy mouth microbiome. In one study,[24] Spivak's team and USDA scientist Kirk Anderson compared the mouth microbiome of honey bees from colonies with or without propolis. The mouth microbiome is a key line of defense against pathogens in honey bees. Bee workers use their mouthparts to feed secretions to larvae, to exchange food with other workers, and to create honey and pollen stores. If the mouthparts become contaminated with disease-causing microbes, these will have opportunities to spread like wildfire. But a healthy microbiome hosts good microbes that keep the bad ones out. The researchers found that bees from propolis-rich colonies had roughly ten times more bacteria in their mouthparts, with many beneficial bacteria and few pathogenic bacteria. So, just like consumption of yogurt or probiotics can shift the human gut microbiome to a beneficial state, propolis can restructure the mouth microbiome of bees by favoring beneficial bacteria that outcompete pathogens.

One of these beneficial bacteria is *Bombella apis*, which occurs widely in hives, honey, and bees. A previous study at the USDA showed that nurse bees transfer these bacteria to larvae as they feed them honey. As a result, these larvae experience less infection when challenged with *Nosema ceranae* parasites[25]. Another beneficial bacterium that occurred at higher numbers in the mouths of bees from propolis-rich colonies is *Lactobacillus kunkeei*. A study at the University of Turku in Finland found that this bacterium produces a protein that stops the growth of *Melissococcus plutonius*, the pathogen that causes European foulbrood.[26] (As a side note, *L. kunkeei* was first described in 1998 as a spoilage organism of wine.[27] Isolated from a Cabernet Sauvignon that was undergoing sluggish alcohol fermentation, researchers described *L. kunkeei* as a novel species that created bad conditions for fermenting yeast cells. This suggests the bacterium could also provide antifungal benefits to honey bees).

RETURN OF THE PROPOLIS ENVELOPE

Clearly, propolis is a highly beneficial substance that works as medicine, allows bees to save on costly immune responses, and gives a boost to their oral microbiome.[28] One major benefit of propolis is that it consists of hundreds of different plant chemicals, making pathogen evolution of resistance much less likely than resistance to the single antibiotic compounds used by beekeepers.[29] Overall, then, propolis could be used as a healthy alternative to the antibiotics and pesticides that beekeepers use routinely in their operations, and to which pathogens are evolving resistance.[30]

When people first domesticated bees, they dragged them out of their natural environments. The tree hollows that wild bees use for their colonies have rough surfaces, holes, and crevices, and bees readily fill them up with propolis. They coat the entire inside of the hollow, producing a beautiful smooth layer, known as the propolis envelope.[31] Now, instead of letting bees build their nests in hollow trees, beekeepers provide them with rectangular hive boxes, built of six smooth planks of wood. As a result, managed bee colonies contain much less propolis than wild colonies, with the unfortunate—and unforeseen—side effect of allowing pathogens and parasites to flourish.

Think of it this way: you are a species that lives in relatively small colonies, foraging on a high diversity of nectar plants, and with a steady supply of medicinal resin. Then, people grab you and force you into high-density boxes kept in huge apiaries in landscapes dominated by single or few agricultural crop plants. And all those nooks and crannies you used to store medicinal resin? Gone. Not only are we forcing impoverished diets on our bees and keeping them at huge densities, increasing the likelihood of disease, we are also taking away one major form of defense against those diseases.

In hive boxes, beekeepers place frames on which the bees build their hexagonal comb cells out of wax, and fill these cells with pollen, honey, or brood (eggs, larvae, and pupae). Beekeepers need easy access to these frames to obtain honey, wax, and pollen and to move bees between colonies. Ideally, hive boxes are easy to open and frames are easy to slide out. Ironically, though, by providing these smooth surfaces, beekeepers force bees to deposit what propolis they still make in the crevices between planks and frames and along the lid at the top of the hive box. And what ends up happening is that bees seal down the frames and seal down the lid. They are putting propolis in all the places where beekeepers want it least. In response, beekeepers tend to select for bees that are less likely to collect resin. That, in combination with the lack of rough surfaces and crevices to deposit propolis, means that modern managed honey bees collect only a fraction of the resin that their wild ancestors did.

To counter that, bee researchers have studied whether we can reintroduce propolis into beekeeping practices. One way is to reselect the behavior. Researchers at the Universidade Estadual Paulista in Brazil chose thirty-six colonies of Africanized honey bees, and picked out the three colonies that produced the most propolis.[32] By crossing queens and drones from these colonies, they bred bees that produced at least five times more propolis. This is yet another example of a behavior that is controlled by genes and therefore can be selected for. Importantly, the researchers found that these colonies also had more honey and pollen.

Spivak thinks such selection is a great idea but may not always be necessary. This is because, despite beekeepers' attempts to get rid of the behavior, collection of resin and production of propolis is second nature to bees: they simply can't resist. One solution is to allow bees to use more resin, not less, but to let them do it in ways that make beekeeping easier. With that idea in mind, Spivak teamed up with a carpenter to produce new hive boxes. Rather than having six smooth planks, the boxes consist of planks with long grooves (think of the grooves you would find in a car tire or in a crinkle-cut potato chip). Moreover, rather than being sanded down, the grooves have rough surfaces, making it easier for propolis to stick. Honey bees readily fill up the grooves and line the boxes with a beautiful propolis envelope that will keep many diseases at bay.

Simone-Finstrom is now a research molecular biologist at the USDA Honey Bee Breeding, Genetics and Physiology Research lab in Baton Rouge, Louisiana. In addition to running programs to select bees for pathogen resistance, Simone-Finstrom is tasked with determining if commercial beekeepers and their bees can benefit from reintroducing propolis. He told me that he and Spivak just finished a two-year study in which they worked together with the world's largest beekeeping operation, a beekeeper in Mississippi who maintains ten thousand bee colonies. The study tracked 120 bee colonies throughout the year. Half of the bee boxes were standard smooth boxes, and half of them consisted of the rough and grooved planks that Spivak developed in collaboration with her carpenter.

They introduced the boxes into the regular beekeeper's operation in Mississippi during the spring of 2019. The operation then moved to South Dakota, where the bees had ample access to wildflowers, from which they collect nectar for honey production. In October, they were trucked to California, where they were kept in holding yards for a few months, and then used for pollination of almond trees. Simone-Finstrom repeated the experiment in 2020. Despite the difficulties of COVID-19 restrictions, which made it harder to collect data, Simone-Finstrom and his team found that the amount of propolis in the rough and grooved boxes doubled in August, and that these boxes had much more propolis than the standard smooth bee boxes. When the researchers checked the same boxes in February, they once again saw the amount of propolis double.

So far so good. But the key question was whether the bee colonies produced as much honey and offspring.

"We always get the question from beekeepers whether greater propolis collection results in less honey production," says Simone-Finstrom. "So, we needed to test that."

Measuring the amount of honey in South Dakota and the number of bees in February, the two-year study showed that the bees in rough and grooved boxes produced just as much honey as those in smooth boxes—provided they have enough nectar sources nearby. More exciting still, they found that rough and grooved boxes contained on average one extra frame of bees, some five thousand extra bees per colony.

Importantly, says Simone-Finstrom, "for pollination services, beekeepers get paid by the number of frames per box. So the extra propolis translates into more money."

As far as agricultural applications of animal medication go, propolis is one of the most promising and concrete examples we have. Not only have numerous scientific studies now demonstrated that this sticky substance is as medicinal to honey bees as it is to humans, but Spivak, Simone-Finstrom, and their colleagues have also demonstrated that a very simple change in the production of bee boxes can reintroduce this medicine in commercial operations—and make the handling of bee boxes easier in the process.

While it may seem daunting to replace all bee boxes at once, Simone-Finstrom does not think that is necessary. "When boxes get damaged," he says, "beekeepers replace them with new ones. So they could replace them with rough grooved boxes one at a time."

He also points out that many beekeepers produce their own bee boxes, and one step in the manufacturing process involves the smoothing of the planks. By taking out that step, beekeepers could simplify the production process, make it cheaper, and allow their bees to deposit more medicinal propolis.

Both Spivak and Simone-Finstrom have high hopes for this application. But the key, they think, is to reintroduce a natural behavior that can help bees and beekeepers alike. What they do not recommend is to apply the propolis in unnatural ways.

"Should we feed propolis to bees?" asks Simone-Finstrom. It is a question he often gets, and some researchers are advocating for it. "Ultimately, propolis is an antibiotic," he says. "It also contains insecticidal compounds."

And feeding those to bees, rather than letting bees take care of their own medication, could ultimately do more harm than good.

11
Dogs Are Dogs

In many ways, our pets are like livestock and honey bees. They mostly live in human-made environments, where we feed them standardized diets and treat them with pharmaceutical drugs. They lack the choices and freedoms that their ancestors had in nature. And just like goats, sheep, cows, and bees, depriving our pets of choices stands in their way of maintaining their own health.

We may believe that our pets are no longer able to medicate themselves. We may think that the thousands of years of domestication have dumbed them down. But just as we saw with livestock and honey bees, dogs and cats, and many other animals that we keep in our care in horse farms and zoos, have retained their abilities to medicate themselves. And in many cases, these behaviors are innate, captured in the genes and the senses that our animals inherited from their wild ancestors.

As we are learning more and more about the medication abilities of our pets, we are also witnessing increased attention to holistic pet health. Just as humans experience the negative effects of lack of movement, limited outdoor activities, and unhealthy highly processed foods, so too are we realizing the negative effects of similar lifestyles we impose on our pets. In the

2021 *New York Times* bestseller *The Forever Dog,*[1] pet health blogger Rodney Habib and veterinarian Karen Becker argue that most of the afflictions that plague our dogs these days are chronic diseases and inflammation caused by unhealthy lifestyles. Many of us feed our pets highly processed foods, expose them to high levels of chemical pollutants, and do not offer them enough outdoor time. Above all, we do not provide our pets with choices. For those of us who own dogs, the picture may look familiar: we determine our dog's diet, we decide where and when to walk, and we tend to stop our dogs from eating dirt or grass. In sum, we make all the decisions for our pets, stripping away their autonomy and life choices.

But it does not have to be that way. Even with some small changes, we can empower our pets to increase their own well-being. We can provide diverse and fresh diets, let our dogs decide where to go on our next walk, create gardens with medicinal herbs—and be a little more accepting of their grass- and dirt-eating habits.

CLAY AND DIRT

Before we discuss the consumption of dirt and grass by our pooches, I need to make a confession: until a few years ago, I was never much of a dog fan. Having grown up with allergies in a pet-free house, I never experienced friendly relationships with dogs. All my childhood memories of dogs are negative: from being jumped on and bitten by escaping German Shepherds and Dobermans on my way to school, to stepping in dog poop during barefoot pick-up soccer games (in the Netherlands of the early 1980s it was not the custom to pick up after your dog).

It was not until I was an adult and my family started occasionally dog-sitting a cute Jack Russell, and I witnessed how much joy he brought to my young children, that I could even imagine inviting a dog into our family. During the COVID-19 pandemic, we finally welcomed a hypoallergenic miniature poodle. We named him Tukkie after "Takkie," a small black dog in the classic book series *Jip & Janneke* by Annie M. G. Schmidt, which many Dutch children have grown up with. Soon after Tukkie joined our family, I started taking him to a small-dog park next to the ballet school where my daughter, Ella, had lessons twice a week. And there I learned some of the interesting behaviors of dogs. One such behavior is the eating of soil. Many dog owners will have observed their dogs digging into the dirt and eating some of it. In that dog park, I also noticed that many dog owners were not happy about this behavior and put an end to it as soon as they could.

Although the expression "eat dirt" is a negative one in the English language, meant to insult someone, the consumption of earth is quite normal—and often healthy. Dogs do it but so do many different species, including elephants, buffaloes, mule deer, tapirs, and many birds and reptiles.[2] Animals also eat clay, chalk, small pebbles, or the mounds of ants and termites. This behavior is known as geophagy (from the Greek words "earth" and "eating"), a term that was first recorded by the ancient Greek philosopher Aristotle in the fourth century BCE.[3] Like animals, many people around the world also consume earth and clay, and its use is so ancient and widespread that scientists have suggested that humans learned to use clay by observing their fellow creatures.[4]

Animals eat earth for several different reasons. In a classic study on clay-licking parrots in Peru, researchers found that the specific clay sought out by parrots neutralizes their highly toxic diet.[5] (This is similar to the goats that eat PEG to neutralize tannins, as we saw in chapter 9). African elephants at Mount

Elgon at the Kenya-Uganda border experience similar benefits from eating clay. These elephants are famous because they eat so much clay that they have excavated caves in the mountainside.[6] The clay that these elephants go after contains kaolin, which neutralizes toxins. Kaolin also coats the intestinal wall, where it stimulates the secretion of mucus, which protects the gut lining from the toxins. These particular elephants do not live in the grassy savannahs that most African elephants inhabit, but make their home in tropical forests, where they consume trees that are loaded with toxic chemicals—hence their hunger for clay.

Geophagy is also common in primates. At least 136 species have been observed to eat soil, clay, rock particles, or termite and ant mounds,[7] including ring-tailed lemurs, chimpanzees, and gorillas. In addition to eating clay for neutralizing plant toxins, primates, including humans, use geophagy to obtain minerals that are otherwise hard to come by.

You may have noticed that these examples of geophagy mostly center on the detoxification of food, which is a change from the focus in this book on parasites and pathogens. As it turns out, however, animals also use geophagy to protect themselves from the harmful effects of infection. In one study, Mary Knezevich of the Caribbean Primate Research Center at the University of Puerto Rico investigated a free-ranging group of 141 rhesus macaques on the Puerto Rican island Cayo Santiago. Almost 90 percent of these monkeys suffered from infections with intestinal parasitic worms and protozoans, but only three of the infected monkeys had noticeable diarrhea. Because almost 80 percent of the macaques eat soil rich in kaolinitic clay, which is known to reduce diarrhea and intestinal upsets, Knezevich concluded that the macaques use the soil to neutralize the effects of parasite infections.[8]

More recently, studies have shown that the consumption of soil can also have major benefits for the gut microbiome, which helps protect against infection. Researchers from Southeast University and Nanjing Xiaozhuang University, both in Nanjing, China, investigated whether consumption of soil by mice can change their gut microbes and thereby enhance their immune system. The researchers were inspired by findings that children who grow up on traditional farms have fewer allergies than children growing up in urban environments, that these children are exposed to more dust and dirt, and that healthy gut microbiomes are necessary for proper immune function. Feeding farm soil to the mice, the researchers found that the animals obtained a greater diversity of microbes in their gut, and that this resulted in a healthier immune system, specifically lowering the immune responses associated with allergies.[9] This is not just because soil is a good source of microbes, but also because soil particles can provide nutrients and microenvironments in which microbes can thrive within the gut.

Soil contains more microbes than any other habitat on our planet, and many animals obtain important microbes from this environment. For example, humans obtain microbes from the soil through direct interaction with soil or through dust or contact with animals.[10] Recent decades have seen a surge in studies that show that healthy and diverse microbiomes are key to animal health. As we saw before, a healthy and diverse mouth microbiome protects honey bees against many different pathogens. Other studies have shown that healthy microbiomes also protect against infectious disease in mice, mosquitoes, bumblebees, locusts, humans, and many other species.[11]

We also know that when the microbiome is disturbed, disease can ensue. This is true, for example, for the severe diarrhea caused by *Clostridioides difficile* (often referred to as *C. diff*). This bacterium can rapidly expand in the human colon when the microbiome is disturbed through antibiotic treatment, causing severe inflammation, fever, bellyache, and diarrhea. One effective treatment is fecal transplant therapy, whereby the feces of a healthy person are transferred to a patient—for example through enema or colonoscopy—to restore the microbiome, suppress *C. diff*, and alleviate disease symptoms.[12]

So, what the last few decades have taught us is that the consumption of soil and clay is common and that it serves many purposes. When your pooch eats dirt, she may do it to neutralize toxins or to obtain essential microbes that she needs for a healthy microbiome and immune function, which will help her against infection with parasites and pathogens. Of course, this does not mean that you should let your dog eat any dirt. Some soils, especially those in urban environments, contain chemical pollutants that could be toxic to your dog. Keeping her away from those would be wise.

GRASS AND PURGE

Since my miniature poodle Tukkie became part of my family, I have joined the vast number of dog owners who have witnessed their dogs eating grass. In contrast with cows and sheep, dogs do not use grass as their basic diet. Yet, they commonly eat the green leaves, and people have witnessed this for a very long time. In his *History of Animals*, written in the fourth century BCE, Aristotle wrote that carnivores "will not eat grass unless they are sick, for some dogs eat grass and vomit it up again, and so are purified."[13] Artistotle's statement is in line with what some of my dog-owning friends have observed, and also resonates with my own observations of Tukkie, who has eaten grass and other plants multiple times. On three occasions, he

FIGURE 11.1. The author's dog, Tukkie, eating grass. Photo by Jakob de Roode.

vomited afterwards. It surely seemed to me that he had stomach upset and used these plants to purge his stomach of whatever was bothering him, thus proving Aristotle right.

Despite Aristotle's statement, however, there is remarkably little scientific evidence that dogs eat grass to vomit. I could only find one report from a scientist in India, who had made observations of twenty-five dogs over the course of two years, and concluded that they usually eat grass to induce vomiting when they have indigestion, liver dysfunction, poisoning, or other gastrointestinal problems;[14] the interpretation was that the coarse structure of the grass, with its bristly leaves, irritates the esophagus and stomach, and thereby induces vomiting.

However, no other study has found strong and clear support for vomiting. One of the most comprehensive studies on grass-eating in dogs was published in 2008.[15] The study consisted of three surveys among dog owners. In the first survey, 25 veterinary students all reported that their dogs ate grass. But only 8 percent noted that their dogs vomited afterwards. The second survey focused on another 47 dog owners. Almost 80 percent of them had noticed their dogs eating plants, mostly grass. But as with the first survey, signs of vomiting were rare. Finally, in a third survey, this one online, the researchers obtained responses from no fewer than 1,571 dog owners. Of these, two-thirds reported that their dogs eat plants on a daily or weekly basis, with the majority of those dogs eating grass. Twenty-two percent of respondents noticed vomiting afterwards. These surveys clearly showed that many dogs eat grass. But because the researchers viewed the numbers of vomiting dogs as low, they questioned whether dogs eat grass to induce vomiting. Personally, I feel that 22 percent is a sizeable proportion of vomiting dogs and am not convinced we can use these studies to totally dismiss the idea that dogs use grass to induce vomiting.

Because the observations of dog owners can be subjective, scientists at the University of New England in Australia took a different approach, and carried out a controlled experiment.[16] The researchers recruited twelve mixed-breed dogs and observed them for three ten-minute sessions per day for six days. The dogs were offered two kinds of grass, potted kikuyu and couch grass, within their kennels and also received a daily kibble meal in the afternoon. The researchers found that all dogs ate grass but that vomiting happened only five times out of the 709 times that dogs ate grass during the study. Dogs ate grass more often and for longer *before* than *after* their kibble meals, suggesting that they ate grass more often on an empty stomach. As with the surveys mentioned above, these

FIGURE 11.2. A dog eating grass. Photo by Radharani / Shutterstock.

researchers concluded that dogs do not eat grass to induce vomiting.

Because relatively few dogs throw up after eating grass, scientists have developed alternative explanations. One hypothesis is that dogs eat grass to remedy digestive problems and illness. So, in the surveys mentioned above, the researchers also asked respondents to note whether they had noticed any signs of illness when their dogs ate grass. Overall, signs of illness were very rare in all three surveys.

The scientists at the University of New England followed up their initial study with another experiment[17] in which they fed twelve beagles a standard diet supplemented with fructo-oligosaccharide. This carbohydrate causes diarrhea and thereby mimics digestive illness. As in their previous experiment, they provided each dog with two types of potted grass, and then monitored grass-eating for several days. Surprisingly, the dogs ate more grass when fed a standard diet than when fed the supplemented diet. This means that dogs without diarrhea ate more grass than those who experienced loose stools, partly contradicting the hypothesis that dogs eat grass in response to digestive problems and illness. I say partly, because diarrhea is only one symptom of digestive illness, and dogs could theoretically respond to other symptoms such as nausea, which the researchers did not study.

Yet another hypothesis is that dogs eat grass for nutrition. Commercial companies promote all sorts of grass-related products as dietary supplements,[18] but most scientists agree that dogs do not eat grass for food. The reason is that dogs cannot actually digest grass. Moreover, they do not generally chew grass when they eat it. Instead, they swallow it whole in big chunks. Sound familiar? You may remember that chimpanzees and other apes, such as bonobos, gorillas, and gibbons, swallow unchewed leaves to dislodge intestinal parasites and purge their guts. Could it be that dogs eat grass for the same reason? Do they eat grass to purge their guts to get rid of parasites? As it turns out, the purging hypothesis seems the most likely explanation for the consumption of grass. And dogs do not just eat grass to get rid of parasites, but also other things that do not sit well in their guts, such as heads from Barbie dolls and Lego pieces.

Humans domesticated dogs from wolves.[19] So, if dogs eat grass to purge their guts, we would expect them to have inherited this behavior from their wolf ancestors. Indeed, in studies dating back to the 1940s and 1960s, researchers found that wolves commonly eat grass, and that scats with grass often contain roundworms or tapeworms.[20] And wolves are not the only carnivores to use grass as a parasite purge. Alan Franck at Florida International University and Arian Farid at the University of South Florida recently dissected hundreds of papers and reports on grass consumption by all known carnivores and came to the same conclusion: when carnivores eat grass, they do it for the same reason as chimpanzees that swallow Velcro-like leaves.[21]

The Carnivora comprise around three hundred different species, such as cats, dogs, bears, civets, genets, and mongooses, most of which eat meat. Of these three-hundred species, only eight rely on plants as part of their dietary needs, including the bamboo-feeding giant and red pandas. Yet, Franck and Farid found that more than one hundred other carnivores eat plants as well, and most of these plants are grasses.[22] They also found that the ingestion of grass is not accidental: in many cases, scats and intestinal tracts contained large quantities. And as we just saw for wolves, Franck and Farid report that in many studies, researchers found scats that contained both worm and grass. This was as true for tigers, cougars, and bobcats as it was for lesser civets and Alaskan brown bears.[23]

The fact that most of these animals eat grass—as opposed to other plants—is also important. Grasses generally have serrations and leaf hairs that are hardened with silica (you may have experienced these structures when trying to pull grass out of the ground and accidentally cutting yourself). So, they are like the rough, hairy, Velcro-like leaves that apes like to swallow. And carnivores specifically seek out particularly coarse grasses, sometimes swallowing grass that is so hairy, tough, and sharp that even goats avoid them.[24] As I already mentioned, most carnivores do not digest grass. In fact, they do not even try to. As the reports assembled by Franck and Farid show, carnivores swallow big chunks of grass or even completely intact blades, sometimes in large bundles, rather than biting off small pieces and chewing them up to aid digestion.

So, as we are learning more about the feeding habits of carnivores, it seems that grass-eating in dogs is a behavior that they obtained from their wolf ancestors, who used it as a way to purge parasites from their guts. While wild wolf ancestors would have experienced more worm infections than modern dogs, the relic grass-eating behavior is not a bad thing. It can still help, as observed by dog owners who have noted the expulsion of worms after their dogs ate grass.[25]

I have focused a lot on dogs here, but I do not want to exclude cat lovers in this narrative. The same researchers who studied grass-eating in dogs by means of surveys among dog owners also surveyed cat owners in 2006 and 2016.[26] Using a few thousand observations, they found that 90 percent of cat owners had seen their cats eat grass. While very few cats showed any signs of illness before eating grass, some 30 percent did vomit afterwards. Thus, vomiting after eating grass is more common in domesticated cats than dogs. This suggests that in addition to the purging of worms—as displayed by wild cats—the consumption of grass may serve an additional function by inducing vomiting. (As I mentioned above, I personally believe that the observation that 22 percent of dogs vomit after eating grass shows that in dogs, as in cats, grass-eating serves multiple functions.)

While we are still learning about the exact function, it certainly seems true that the consumption of grass by our four-footed, furry friends is not harmful. In fact, because of the potential health benefits, pet owners would do well to let their pets turn into temporary grazers when they feel like it. But there is a catch: while the consumption of grass itself is not harmful, dogs and cats can become severely ill from pesticides applied to lawns.

And they may contract parasites from eating grass that has been contaminated by other animals. As such, the key is to find some healthy and clean grass, whether in one own's backyard or in an indoor grass garden.

SENSORY GARDENS

Letting dogs be dogs and cats be cats, allowing them to eat dirt and grass when they need it, is but one way in which we can offer our pets some options in their medicine cabinet. There is a growing interest in holistic approaches to maintain and increase pet health. In *The Forever Dog*, authors Habib and Becker describe how a shift from ultra-processed kibble to natural foods can reduce chronic disease, increase health, and extend a dog's lifespan.[27]

Veterinarians are also interested in reapplying traditional medicine to improve our pets' health. In the Netherlands, the Institute for Ethnobotany and Zoopharmacognosy has provided advice on what herbal medicines to use in domesticated animals.[28] And in her book *Psychoactive Herbs in Veterinary Behavior Medicine*, veterinarian Stefanie Schwartz summarizes the scientific evidence for the medicinal properties for a wide range of herbs from traditional Western, Ayurvedic, Native American, and Chinese medicine.[29] Based on this evidence, she develops recommendations for their use as a supplement to modern veterinary care.

Another way to provide animals with healing herbs is by developing medicinal gardens in which our animals can find their own medicines, much like the foodscapes that Juan Villalba from Utah State University envisions for livestock. The Bath Cats & Dogs Home in Bath, England, for example, developed what they call a sensory garden. It has a diversity of medicinal plants and contains a tunnel made of birch, where arthritic dogs hang out and nibble at it. The garden also provides clay that animals can use to treat wounds and to fight fungal infections.

The idea to include clay came from Caroline Ingraham, a UK-based consultant and author who developed methods to let animals use minerals, plants, and plant oils as self-medication.[30] Coining these methods "Applied Zoopharmacognosy," Ingraham has tried to raise awareness about animal self-medication among vets, zoos, sanctuaries, and pet owners since the early 1990s. As you may remember from chapter 1, "zoopharmacognosy" is the term that scientists coined to describe self-medication behaviors of animals in the wild. Adding the word "applied," Ingraham developed a new methodology by which we can reintroduce self-medication to our pets and other companion animals.[31] The methodology consists of giving animals choices between different essential oils, herbs, and minerals.

As one demonstration of the approach, Ingraham told me about the time she spent at the Sheldrick Trust in Kenya, a charity that cares for orphaned elephants. In 2007, two local Maasai boys found an eight-month-old elephant stuck in a water well dug for livestock. The steep slopes had prevented her from exiting, hyenas had attacked her trunk, and she had a massive wound on her back, close to the spine. The well water was putrid with the decaying remains of animals, and the elephant was suffering from bacterial infection. The baby elephant made it to the Sheldrick Trust orphanage by way of light aircraft. When she arrived, she was named Sinya, and unfortunately she was in bad shape. Cleaning her wounds required physical restraint, and she was not responding to aggressive antibiotic treatment. The veterinarians were concerned that her infection would cause septicemia, and they asked Ingraham for assistance.

When Ingraham arrived at the orphanage, she encountered a severely sick and depressed elephant baby.

"I asked myself how she would behave in the wild," says Ingraham.[32] "Maybe she would cover her wounds in clay. So I got a lot of green clay and covered her back."

The idea to use clay made a lot of sense. People have used clay for probably as long as they have roamed the earth, either internally, where the clay minerals can adsorb toxins, bacteria, and viruses in the gut, or externally by rubbing clay into wounds.[33] There is evidence for the use of clay for wound treatment by *Homo erectus*, *Homo neanderthalensis*, and ancient Egyptians, Greeks, and Romans.[34]

The clay chosen by Ingraham is often used for cosmetic face masks and is green because it contains iron and decomposed algae. Research has shown that green clay is an effective treatment of Buruli ulcer, which is an infection caused by the flesh-eating bacterium *Mycobacterium ulcerans*. This pathogen is related to the bacteria that cause leprosy and tuberculosis, and destroys fat tissues under the skin, mostly on arms and legs, leading to severe tissue damage and often requiring surgery or amputation. Impressively, treatment with green clay not only helps kill the bacteria because of its antimicrobial minerals[35] but also restores tissues.[36] And so it did for Sinya, whose wound started healing within eleven days. The Kenya Wildlife Trust now routinely uses green clay as a first-line treatment.

In addition to treating Sinya's wounds with clay, Ingraham found that the elephant chose to consume garlic and clove oils when offered different herbs and essential oils. People have used garlic as a medicinal plant for thousands of years. The essential oils of garlic contain chemicals that interfere with protein production in bacteria and disrupt their cell walls and membranes. The chemicals also prevent the formation of biofilms, by which bacteria clump together, cause disease, and make it harder for antibiotics to do their job.[37] Clove essential oils similarly have compounds that break down bacterial cell walls and interfere with the synthesis of bacterial DNA, thereby preventing bacterial reproduction.[38] Lab experiments have even shown that the essential oils of garlic and clove are more effective than antibiotics at killing the bacteria that cause persistent Lyme disease.[39]

Despite the growing popularity of providing animals with natural foods and opportunities to self-medicate, however, there remains skepticism as well. Some of that goes back to the idea we saw before: either that animals cannot medicate themselves in the first place or that domesticated animals have lost these abilities as they were bred and dumbed down from their wild ancestors. Some of the skepticism regarding the use of essential oils, such as garlic and clove oils, stems from the misuse of these oils and their association with pseudo- and antiscience. Robust science remains lacking, and many veterinarians would like to see controlled experiments to show that the methods are effective—and safe. The latter is especially important: as we have seen, many drugs have bad side effects, and when used in the wrong dose, they can do more harm than good. We may even end up killing our beloved friends.

This is why Carly Hillier, director of Whitethorn Equine Health in Lavagh, Ireland, is teaming up with research scientists. Hillier studied equine science at University of Limerick and became interested in applied zoopharmacognosy when one of her horses experienced ongoing colic. She told me that she has used applied zoopharmacognosy to help many a horse cure itself from anxiety, gut infection, and abscesses.

But despite those successes, she says, "it is important that we back up our experiences with actual data that is indisputable."[40]

This is why she has expanded the team at Whitethorn Equine Health to include a research scientist who is carrying out controlled experiments to document the safety and efficacy of applied zoopharmacognosy.

I agree. We do need well-designed experiments to demonstrate that these applied methods of animal medication are effective and safe. Once that is done, I believe that the applications will be far-reaching: from curing skin infections in our pets to reducing anxiety in zoo animals.

In the meantime, we can provide paddocks, gardens, and enclosures that have a wide variety of medicinal plants. Growing sensory gardens is not restricted to animal shelters or zoos. Most people who own pets can create such a garden themselves, either by repurposing some of their lawn or by planting herbs in pots on their balconies. In doing so, we can help our pets live healthier lifestyles and give them back some of the free choices their ancestors enjoyed in the wild.

But we can do a lot more with the evolved medicinal wisdom of animals. As we'll see in the next chapters, animals can also help us discover new drugs to reduce human suffering.

12

Elephant Educators

The use of drugs is perhaps the most defining common feature of human medicine. In traditional Chinese medicine, doctors use herbs and teas as one way to find balance between the inseparable cosmic forces of yin and yang. In the traditional Indian medicine system, Ayurveda, herbal treatments are combined with diet, meditation, and yoga to find balance between soul and body. Ancient Mesopotamians employed plants such as poppy, saffron, garlic, and juniper as drugs, and produced the oldest written medical records on cuneiform clay tablets between roughly 3500 BCE and 500 BCE. Medical Egyptian papyri list close to nine hundred prescriptions based on five hundred different substances, including tannin and willow bark. In European medieval times, monasteries maintained medicinal gardens, filled with an assortment of healing plants[1].

While ancient civilizations did not extract their drugs into the single compounds that modern-day pharmacologists develop into smooth pills, they did recognize the power of plants and other natural resources—such as the polypore mushrooms used by Ötzi the Iceman. Long before people understood what chemicals are, they used extracts to cure themselves of illness. And long before people knew about parasites and pathogens, they used natural products to kill them or relieve the symptoms of disease. While some modern scientists may look down on traditional practices, it is good to remember that many of the drugs that medical doctors prescribe to their patients derive from nature and have been used by traditional healers for thousands of years. Indeed, until 150 years ago, almost all drugs were derived from plants.[2] Even today, plants provide the basis for the majority of drugs—either by providing chemicals that are directly used as a drug or by providing chemicals on which new drugs are modeled.[3]

While many of us now use purified chemicals as drugs, it is important to realize that traditional medicine is not a thing of the past. With only a fraction of the human world population having access to Western medicine, most people living today still depend on—or choose to use—some form of traditional medicine.[4] And much of that traditional medicine relies on plants. As a result, medicinal plants provide the primary form of medicine for more than 70 percent of people living in most developing countries. That is a staggering 4 billion people overall.[5]

Because of their long-term use, traditional medicines are generally considered safe to use.[6] Moreover, the use of natural products—as opposed to purified chemical drugs—could have additional benefits, by slowing down the evolution of drug resistance. A case in point: the antimalarial quinine, obtained from the bark of the Cinchona tree, remained effective for hundreds of years, whereas the industrially produced chloroquine lost its effectiveness within twenty years because malaria parasites evolved resistance to the drug.[7]

Despite these benefits, regulation of herbal treatments is often lacking, which could result in inappropriate dosing and harmful side effects. This is why efforts are underway to better understand the medicinal effects of herbal treatments and to use them more effectively. For example, the Natural Chemotherapeutics Research Institute in Kampala, part of the Ugandan Ministry of Health, is tasked with evaluating medicinal plants to improve treatment efficacy and safety.[8]

Humans have used nature's medicines for as long as we have roamed the Earth. With the continued and renewed interest in traditional medicine, we'll use this chapter to explore how animals have helped humans develop treatments based on herbs and other natural substances. We will then use the next chapter to ask how animals can help the pharmaceutical industry produce modern drugs and insect repellents.

DANCING GOATS

People have long looked at animals to gain medicinal wisdom.[9] According to legend, after witnessing his goats consume berries of a coffee bush in the Kaffa region in what is now Ethiopia, a herder named Kaldi noted that the goats became invigorated and frisky. This led an abbot of a nearby monastery to roast the berries and brew them into a drink that many of us are now dependent on (you may need it to get to the end of this book).[10] The use of cocaine finds its roots in Peruvian llamas who ingest leaves of *Erythroxylum coca* when moving down to lower elevations of the Andes to improve their stamina.[11] And as I mentioned before, Native American healers and shamans credit bears for being nature's master of medicine. When waking up from their monthslong hibernation, bears consume the bark of willow trees, which reduces aches and inflammation. The bark contains salicylic acid, the chemical that forms the basis of aspirin.[12]

You may have heard of "horny goat weed," an herb in the genus *Epimedium*. Legend has it that a Chinese goat herder noticed that goats eating this herb became, well, you can guess.[13]

Laboratory studies have shown that when rats feed on icariin, a flavonoid chemical from this herb, they produce more sperm; and castrated rats even manage to have proper erections.[14] The herb was recorded as a medical treatment in the Chinese medical classic *Shen Nong Ben Cao Jing* four hundred years ago, and also featured in the most famous Chinese medicine document *Ben Cao Gang Mu*.[15] In China, Japan, and Korea, people use horny goat weed to treat impotence, as well as other ailments such as forgetfulness and osteoporosis. While traditional medicine relied on adding the leaves of the herb to wine or water, these days the weed is also available in the form of tablets and noodles.[16]

We have met Mohamedi Seifu Kalunde, who worked for many years with Michael A. Huffman to study self-medication in chimpanzees in Tanzania. He and his family observed and mimicked porcupines, elephants, wild boar, chimpanzees, and other species to develop new treatments for diarrhea and bacterial infections. Mohamedi's grandfather, Babu Kalunde, also developed an herbal treatment by watching an elephant. As the story goes, he observed an elephant with an upset stomach mixing lightly chewed leaves of a tree called *Piliostigma thonningii* with water in his pharyngeal pouch, an organ used to store water at the base of the elephant's tongue. The elephant then swallowed the water and discarded the leaves. Babu Kalunde started using water extracts to treat stomach upset in people. The plant is known as "Munyonga nTembo" by the WaTongwe, which can be loosely translated as "elephant pulling out of the ground." Today, the plant, which has demonstrated antibacterial activity, is used across Africa to treat many afflictions, including malaria, fever, diarrhea, and respiratory infections.

On the island Pohnpei in the Federated States of Micronesia, traditional healers discovered new medicines by observing the plants eaten by sick animals.[17] In Gabon, Africa, people started using *Tabernanthe iboga* plants for religious rites when witnessing gorillas, boars, and porcupines dig up and eat the roots, and then go into wild frenzies.[18] And in Laos, people have long observed that elephant calves eat their mother's feces at the time of weaning. Because elephants eat many medicinal plants, people view the dung essentially as a medicinal cocktail. As such, some people make a concoction of feces, then drink it to treat diarrhea.[19] Given our insights that fecal transplants can restore the microbiome as discussed earlier, it is well possible that this treatment acts in a similar way.

People not only look at animals to develop medicinal treatments for themselves but also for their domesticated and companion animals. Just as most people in the world do not have ready access to Western medicine, so most animals managed by humans do not have access to Western veterinary drugs, whether that's because of nomadism, remoteness, or because many animals have not yet been studied.[20] As such, much of the veterinary care around the world builds on herbal and ethnoveterinary medicine, and the knowledge for these often comes from observing wild animals. In Central Karakorum National Park in Pakistan, hunters and shepherds believe that wildlife, including house sparrows, Himalayan ibexes, and flare-horned markhors, use berries, leaves, and bark of *Berberis* plants as self-medication. Therefore, when domestic animals, such as sheep, horses, cows, or donkeys, are sick, they take them to those plants or bring the plants to the animals if they are too ill to move.[21]

Similarly, in the Karamoja Region in Uganda, traditional healers and pastoralists have observed some fifty self-medicating behaviors in their goats, cattle, sheep, donkeys, camels, poultry, and dogs. The healers and pastoralists use most of these treatments for their animals, as well as their families, suggesting they mimic the medication behaviors of animals for the well-being of their animals and themselves.[22]

In Southeast Asia, humans and elephants have lived in close relationships for thousands of years. In Laos, for example, people have traditionally used elephants for transport and assistance with logging. The elephants are not domesticated, but because they spend a good chunk of the year living in villages with their caregivers (mahouts), they can also not be considered wild. Mahouts let the elephants forage on their own: the elephants are either let free to roam, or mahouts will tether them on long chains and move them between different locations so that they can choose their own diets. In one study,[23] researchers interviewed sixty mahouts to learn about elephant health and medication. Many mahouts reported diarrhea and constipation, as well as viral infection by elephant endotheliotropic herpesvirus. Mahouts noted over sixty associations between specific health conditions and the consumption of particular plants, and found that the elephants recovered from their health conditions when using these plants.

As with the livestock herders in Pakistan and Uganda, mahouts told the researchers that they bring the elephants to areas where they know they'll have access to the plants when they need them. Some mahouts will collect plant roots and give them to the elephants as an actual treatment: one mahout would crush the root and stem of a particular vine and mix it with salt to feed it to the elephants. And they don't just use the plants for elephant health. Of the twenty medicinal plants used by elephants, mahouts use fifteen for therapy in their own households. One mahout noticed an elephant consuming bark when having diarrhea, and now uses the same bark for himself and his family. Another collects roots used by elephants and macerates them in alcohol to create a healing tonic. While

FIGURE 12.1. The mahout (caregiver) Oupe offering a dietary supplement to an elephant. Photo by Paul G. Keil, Assam, January 2015.

scientific studies have provided evidence for the medicinal uses of many of these plants, researchers did note that for one plant named *Lithocarpus auriculatus*, no medicinal reports yet exist. This means that even today, elephants continue to discover medicinal plants unknown to us.

As in Laos, mahouts in Thailand have also learned from the elephants themselves which plants to use in veterinary care.[24] And amazingly, the close association between elephants and mahouts has not just resulted in the transfer of medicinal knowledge from elephants to humans, but also from humans back to elephants.[25] Thus, when a mahout notes a specific ailment in an elephant, they will take that elephant to a specific plant remedy—even if the elephant has never used it itself. This allows the elephant to sample the plant, so it can learn to use the plant in the future. In this sense, mahouts have taken over the role of other elephants in providing the social interactions needed for developing self-medication behaviors.

SACRED SECRETS

So, it is clear that humans have long looked at animals to develop medicinal treatments. What is not necessarily clear is quite how much human medicine derives from animal knowledge. For some researchers, the answer is obvious. When I asked Michael A. Huffman how many treatments he believes were at some point derived from copying animal behavior, he answered by turning around the question: "How many are *not* based on animal knowledge?"

I myself am convinced that much medicinal knowledge stems from animals. But to some extent, the answer to the question of how much knowledge we obtained from animals remains elusive. The reason is that many traditional healers do not openly share their knowledge. In most cultures, the information on healing is passed on orally, and often only to a select few people within communities. The knowledge that healers obtain is often viewed as exclusive and sacred, and divulging those secrets to scientists—or even nonhealers within their own communities—is not necessarily done.

Moreover, some healers may simply not want to admit that their knowledge stems from observing animals. In the study of healers and pastoralists in Uganda that I described earlier, two healers who admitted that they had learned from animals were laughed at, suggesting that some healers may not be so eager to share this information. In that particular study, healers preferred to credit deities and people, whether dead or alive, with their medicinal knowledge[26].

Interestingly, traditional healers not only consider their own knowledge secret, but also that of animals. For many of them, it takes trust and spiritual connection to learn. As one Native American healer told Huffman during an ethnomedicinal conference: "You were listening and she [Chausiku] showed you her secrets."

13

Cats and Catnip

Much healing knowledge is sacred (and secret), but that which we have learned has made a great impact on Western medicine. Many of the examples of medicine copied from animals described in the last chapter involve the use of herbal treatments—as opposed to the use of purified chemicals in neat pills. Because most people in the world still rely on traditional medicine, animals have played a crucial part in the development of human medicine. But the impact of animals is not restricted to traditional medicine. Indeed, herbal treatments also provide the basis for the pharmaceutical industry to develop single-compound drugs.

For hundreds of years, scientists have turned to traditional healers to identify plants and other natural products that can serve as the source for the chemicals that make up neat and tidy pills. This is not surprising because the hunt for drugs is so much easier when we know where to look.

The development of modern drugs is an arduous process.[1] It can take many years, and hundreds of millions of dollars, to produce a chemical that can be sold as a neat tablet. These days, the drug discovery process often starts with the screening of a large library of hundreds to thousands of compounds. Say the goal is to find a drug that kills bacteria. Drug developers will then add each compound to a culture of bacteria to see if it reduces their growth or kills them. They will also test the compounds in animals and in human cell cultures to see if the compounds have toxic side effects. If the drugs don't, the scientists next need to test the compound in a human trial to make sure the drug works and is safe. And if that is the case, the compound needs to be approved by governmental agencies before it can be brought to market. Clearly, finding a drug among thousands of random chemicals is much harder than focusing on a small set of chemicals already known to include medicinal compounds. This is where traditional knowledge comes in.

Take quinine. As early as the sixteenth century, Spanish conquistadors were introduced to Peruvian or Jesuit's bark by the Quechua people in Peru. The Quechua obtained the bark from the Cinchona tree—which they called the fever tree—and brewed it into a bitter tea to treat malaria fever. Jesuit priests brought it back to Europe. The active ingredient, quinine, was later isolated by French apothecaries in 1820.

We encountered digoxin in chapter 7. This chemical was first isolated from purple foxglove (*Digitalis purpurea*) in 1930 and is commonly used to treat abnormal heart rhythms and heart failure. The plant was prescribed by English healers as early as the eighteenth century, but has probably been used for thousands of years to treat pulmonary edema: the swelling of body parts that results from the heart failing to pump blood around the body (also known as congestive heart failure).

And then there is aspirin, which we have seen many times throughout this book. An Egyptian papyrus roll from around 1534 BCE mentions willow bark to relieve inflammation, and the Greek philosopher Hippocrates (often viewed as the father of modern Western medicine) recommended it to relieve pain and reduce fever. English Reverend Edward Stone rediscovered the use of willow bark as a fever reducer in 1763, and in 1838, the Italian chemist Raffaele Piria obtained a more potent form of willow extract, known as salicylic acid, and named after the willow genus name *Salix.*[2] The German pharmaceutical company Bayer eventually improved the chemical to reduce side effects and created the wonder drug aspirin. In all these cases, drug developers did not have to stumble in the dark: they could focus on specific plants that were already known to contain medicinal compounds.

BARK, ROOTS, LEAVES

Studying how traditional cultures use plants for medicine to develop drugs is not a thing of the past. My colleague Cassandra Quave, a professor of dermatology and human health at Emory University, and author of *The Plant Hunter*,[3] focuses on discovering new drugs based on the careful and detailed documentation of knowledge of traditional healers. Quave has told me that scientists have exhaustively studied only a few hundred plant species for medicinal compounds.[4]

"But we have over thirty-three thousand species that are used in different forms of traditional medicine," she says.

That is about 9 percent of all plant life, and leaves plenty of opportunities to discover new drugs.

Much of Quave's work focuses on treatments of skin conditions, and in particular those caused by *Staphylococcus aureus*, or "staph" for short. In interviews with Italians and Arbëreshë, an ethnic group that moved from Albania to Italy, Quave has found that many people in these communities use the sweet chestnut tree (*Castanea sativa*) to treat a wide variety of skin conditions. Local healers use chestnut leaves to produce teas to apply to skin that is irritated, inflamed, or affected by eczema. As she describes in *The Plant Hunter*, Quave collected leaves and started extracting different chemical fractions from chestnut leaves. She then exposed staph bacteria to these fractions. One of them, which she refers to as fraction 224C, did not kill the bacteria but was able to disrupt staph's production of toxins. Many bacteria, including staph, work together in a process called quorum sensing. For staph bacteria, this means sending chemical messages to each other to collectively produce toxins that rupture red blood cells and damage cells and tissues, so as to provide a suitable source of food. As Quave found, extract 224C interfered with quorum sensing. And so it reduced the production of toxins.

In another study, Quave found that an extract from blackberries inhibits staph biofilms. Biofilms are blankets of cells that bacteria form in wounds and tissues but also in medical equipment, such as intravenous (IV) tubes that doctors use to deliver drugs or fluids to the patient's blood stream. They are especially difficult to treat, as they often become refractory to even the best antibiotics we have. But Quave showed that in combination with antibiotics, the blackberry extract broke up the biofilms, such that fewer cells ended up sticking to IV tubing.

Humans are not the only species with traditional medicinal knowledge. As we have seen throughout this book, animals possess much wisdom about medicine. This is why researchers are now increasingly turning to them to identify products that could be the source of the next wonder drug.[5] The premise is that by looking at animals, we can make the task of finding potential sources of chemical drugs from the vast natural pharmacy much easier.

So far, the study of prophylaxis in chimpanzees in Kibale has resulted in the discovery of two new antimalarial compounds in *Trichilia rubescens* plants.[6] And laboratories in the United States and Malaysia have found anti-tumor effects of vernodalin, a compound from the bitter plant *Vernonia* used by chimp Chausiku.[7] A patent for its application to treat breast cancer has been awarded.

In 2022, I had the opportunity to talk to Ulrich Maloueki,[8] a researcher at the University of Kinshasa in the Democratic Republic of Congo. Maloueki has been working in the lab of Désiré Musuyu Muganza, an expert in plant chemistry. Maloueki is interested in the feeding behaviors of primates and has studied gorillas, bonobos, and chimpanzees. In his field work he is especially interested in the use of food items that appear unusual.

"The animals like fruits," he says, "so I look for animals eating bark, roots, and leaves."

Accompanied by field guides, Maloueki always asks them whether they are familiar with the plants used by the animals, and he has found out that many of these plants are used as medicine in their local communities. He will then check the literature to find out what is known about the plants. He also obtains chemical extracts, which the lab tests against malaria parasites and bacteria.

Using this approach, Maloueki has already found that bonobos swallow the juice from the stems of *Megaphrynium macrostachyum* and *Palisota hirsuta,* and that extracts from the plants can reduce both the growth of malaria parasites and bacteria in the lab.

"The next step," he says, "is to identify the specific chemicals that are responsible for the antimalarial and antibacterial activity."

Maloueki has a clear long-term goal: to develop the antimalarial and antibacterial chemicals into drugs for humans.

Maloueki is not the only scientist studying animals with the goal to ultimately help humans. As we will see in our final journey, researchers in Japan have turned to cats to try to control the devastating effects of six-legged blood suckers.

RUB 'N' ROLL

Mosquitoes form one of the biggest scourges of human suffering in the world.[9] Not only do these insects bite people at will, they also transmit many diseases, including malaria, yellow fever, and dengue, thereby sickening and killing many millions of people around the world each year. It is no wonder that controlling mosquitoes and the diseases they transmit is a major priority of biomedical research.

While many researchers are focusing on developing drugs and vaccines to prevent infections and disease, others are trying to develop chemicals that can effectively kill or deter mosquitoes. Interestingly, one promising chemical deterrent has been used by cats for millions of years—and researchers are now hoping to put that cat knowledge to good use.

To understand how cats are helping us in the fight against blood-sucking mosquitoes, we need to go to Iwate University in Japan to meet a student named Reiko Uenoyama and a professor named Masao Miyazaki. Uenoyama had owned and loved cats ever since she was a small child. So, when she entered Iwate University as an undergraduate student in 2016, she was elated to find out about the lab of Miyazaki, who studies cat behavior. When it was time to select a research lab during her junior year, the choice was easy.

"Cats have many strange behaviors," she says.[10] "But I was especially interested in their response to catnip and silver vine."

Many a cat owner will recognize what is called the catnip response: when smelling the plants catnip (*Nepeta cataria*) or silver vine (*Actinidia polygama*, also known as matatabi in Japanese), domestic cats will start rolling over and rubbing the leaves into their face and fur. They will lick the leaves and bite them—but not eat them. The rolling, rubbing, licking, and biting may continue for some five to fifteen minutes. To an observer it may seem that the cat is experiencing euphoria. People have known about the catnip response for hundreds of years: a Japanese botanist described the behavior as early as 1704, and Japanese folk stories, drawn up in pictures by Tsukioka Yoshitoshi in 1859, tell of mice using silver vine as a weapon to intoxicate cats.[11] Yet, why exactly cats frantically rub the plants had remained a mystery.

"The cat is a mysterious animal," says Miyazaki. "Cats are very popular, but no one really understands their physiology and their behavior."

Ironically, people have used catnip as a medicinal plant and as an insect repellent for a long time,[12] with remains from a Neolithic village showing evidence for cultivation of catnip since at least 3900 BCE.[13] In the 1940s and 1950s, Japanese and American chemists identified a number of iridoid chemicals in catnip and silver vine.[14] Some of these iridoids are potent pest repellents, and the plants produce them to avoid being eaten by insect herbivores.[15] Yet, despite the clear insecticidal effects of the plants and their chemicals, no one seemed to think that cats use the plants for the same reason. The story is reminiscent of honey bee resin, as we saw in chapter 10: despite the common use of resin as medicine by humans, no one seemed to think that bees could use the sticky substance as medicine themselves.

FIGURE 13.1. Gray cat enjoying fresh catnip. Photo by Anna Hoychuk / Shutterstock.

That all changed when, in 2012, Toshio Nishikawa, a professor of organic chemistry at Nagoya University, contacted Miyazaki. Nishikawa had been interested in the catnip response since he learned about it in high school. At Nagoya, he synthesized several iridoids in his laboratory and was looking for collaborators who were experts in cat behavior. At the same time, Miyazaki had become interested in studying the catnip response but was not able to commercially obtain iridoids for his experiments. A new collaboration was born, and Nishikawa and Miyazaki started working together in 2013.

They first decided to do a thorough chemical analysis of silver vine. Catnip had been researched well before, and chemists had identified their main iridoid as nepetalactone. Japanese chemists had also isolated chemicals from silver vine, but they did not yet know which of these chemicals make the plant attractive to cats. Miyazaki started extracting the chemicals of silver vine leaves, then separated them into six fractions, each containing different mixes of compounds. Next, the researchers added the chemical fractions to filter papers, which they attached to the floor of experimental cages. They also placed control filter papers—without the chemicals—on the floor. Then they released a cat into the cage to see how it would respond to the different fractions.

As it turned out, most of the fractions did not send cats into a frenzy, but one elicited a very strong response. The cats rolled around and rubbed their faces on the filter paper—just as they would with actual silver vine leaves. Further analysis suggested that the main compound in this fraction was a chemical named nepetalactol. Nishikawa synthesized nepetalactol and then Miyazaki tested it in cats. When adding this single compound to filter papers, it elicited a similar strong response from the cats, demonstrating that this chemical on its own is a big reason that cats go frantic over silver vine.[16]

A cat's reaction to catnip and silver vine has often been described as extreme pleasure. So, when Uenoyama joined the lab in 2018, Miyazaki suggested exploring the μ-opioid system, the hormonal system that controls feelings of reward and euphoric effects in mammals. When measuring blood levels of hormones, they found that one type of endorphin (an opiate made by the body) increased in concentration after cats were exposed to nepetalactol. They then injected six cats with a chemical that blocks the receptors that normally bind the endorphins and set in motion a response that results in extreme pleasure and euphoria. As a result, the cats stopped rubbing their faces and rolling over the paper, despite smelling the otherwise pleasure-inducing molecule.[17] Thus, the bodies of cats respond to the smell of catnip and silver vine by producing pleasure-invoking hormones, which keep the cats rolling, rubbing, biting, and licking.

To determine if the single chemical, nepetalactol, also elicited typical responses in other cats, Uenoyama and Miyazaki persuaded collaborators at the Osaka Municipal Tennoji Zoological Gardens and Kobe Oji Zoo to run their tests with an Amur leopard, two jaguars, and two Eurasian lynx. Similar to their domestic counterparts, all of these big cats happily rubbed their faces on and rolled around on the filter papers with nepetalactol. Not that it was easy to figure this out.

"It was very difficult," says Uenoyama. "In a normal cat we can do the experiment multiple times. Previous studies have shown that there are no toxic effects in catnip and silver vine, but some zookeepers are afraid of that. So, we could do the experiment only once or twice."

Given that domestic cats and wild big cats share a common ancestor that lived around 10 million years ago, Uenoyama and Miyazaki's results suggest that the response to nepetalactol evolved early on. They therefore hypothesized that the catnip response was important for the cats' survival and was selected for in early cat evolution. In 2020, Miyazaki organized an international conference at Iwate University. He invited Jane Hurst, a professor of mammalian behavior and evolution at the University of Liverpool, to speak at the conference. As they were discussing different research projects, and watching many videos of cats having fun with silver vine, Hurst suggested to Miyazaki that the cats may rub and roll to coat themselves with the insect-repellent iridoids.

At this point, says Miyazaki, they started wondering if cats use catnip and silver vine as protection against mosquitoes. This would certainly provide a survival advantage to the wild ancestors of our domestic cats: in the wild, the blood-thirsty insects not only suck feline blood, they also transmit many different parasites, including heartworms that cause infections of the lungs and heart.

Showing that the catnip response is a protection against mosquitoes required a few pieces of evidence. First, the researchers had to show that the chemicals actually repel mosquitoes. To test that, Uenoyama and Miyazaki placed mosquitoes in cages and provided them with shelters they could fly into and away from any smells they didn't like. When exposed to silver vine leaves or nepetalactol, many more mosquitoes flew into the shelters, showing that the chemical indeed deters the insects.

Second, if the cats rub and roll to repel mosquitoes, the behavior should actually result in the cats coating themselves with chemicals. To test that, Uenoyama and Miyazaki set up cats in cages where they could touch the papers with their faces but not roll on them: so, instead of placing the papers on the floor, they attached them to the walls and ceilings of cages. It turned out that tested cats in the experiment rubbed more on papers with the chemical than on control papers. But they no longer rolled on the floor of the cage, which suggests they roll specifically to make contact with the chemicals rather than rolling around for fun alone.

The researchers then had to demonstrate that the cats contract enough chemicals to repel mosquitoes. This was hard to test because the chemicals on the fur of cats are so diluted that they were below the detection levels of the researchers' equipment. Working around this problem, the researchers decided to rely on cats' noses, which are better at detecting chemicals than highly advanced human-made machines. For the cats that had rubbed their faces and heads on papers with nepetalactol, the researchers took a clean towel and wiped it along the cats' heads to transfer the chemicals (if any) to the towel. They then presented those towels to new cats. As expected, the new cats went into a frenzy when exposed to the towels, showing that enough nepetalactol was transferred from the paper to the original cats' fur.

The final piece of the puzzle was then to show that nepetalactol protects cats from mosquito bites. To test this, Uenoyama and Miyazaki used six pairs of cats: they applied nepetalactol to one cat's face but not the other, and then placed the cats on opposite sides of a cage. Adding mosquitoes, they found that cats with nepetalactol received half the number of mosquito bites as the cats without the chemical. They found similar results when testing mosquitoes on pairs of cats, one of which had rubbed its head on silvervine and one of which had not. Thus, Uenoyama and Miyazaki's experiments showed that cats specifically make contact with silver vine to transfer iridoids to their fur, and that this results in fewer mosquito bites.[18]

But an important question remained. Although the experiments showed that rubbing and rolling allows cats to coat themselves in mosquito repellent, the experiments did not explain why the cats lick and bite the leaves as well. The cats hardly ever eat the plants, so they must bite and lick for a different reason. To understand that part of the puzzle, Uenoyama and Miyazaki decided to compare the chemicals in intact and cat-damaged leaves.[19] They found that damaged leaves emitted up to ten or twenty times more iridoids than undamaged leaves. For those of us who like to cook with fresh herbs, this may not be too surprising: breaking the leaves of mint or rosemary similarly results in stronger smells. What was surprising, however, is that silver vine leaves did not just emit more chemicals, they also produced more. Somehow, the licking and biting set in motion a physiological process by which the leaves produced more iridoids.

The biting and licking also changed the relative concentrations of different types of iridoids. Where nepetalactol was pretty much the only iridoid in undamaged leaves, damaged leaves produced additional iridoids. Interestingly, when exposed to this iridoid mixture, cats rubbed and rolled for even longer than when exposed to nepetalactol alone. And importantly—the chemical mix from damaged plants also was much more repellent to mosquitoes. Thus, cats not only use silver vine as mosquito repellents, they also actively increase the potency of the chemicals by licking and biting the leaves. Cats are the masters of their own medicine—much like bumblebees that change the chemicals in medicinal nectar[20] and wood ants that add formic acid to resin to make a more potent antimicrobial.[21]

Not that the cats need to know anything about this. The evolution of the catnip response is driven by pleasure-evoking physiological processes that do not require learning or higher cognition.

"It is innate behavior," says Uenoyama. "The cats may not know that the chemical repels mosquitoes, but they respond when they sniff the compound."

The pleasure experienced by cats is not much different from the stronger taste experienced by woolly bear caterpillars that eat food with more alkaloids. And, as a study of a pedigree of Siamese cats showed back in 1962, the catnip response is likely driven by a single gene, which supports the idea that the behavior can be easily passed from parents to offspring.[22]

People have used catnip, and its main iridoid nepetalactone, as a mosquito repellent. But Uenoyama and Miyazaki's research showed that with silver vine's nepetalactol, cats have discovered a much more potent chemical. And that raises the question of whether it may be useful for humans as well.

So, the researchers decided to test the chemicals on themselves.

"We felt so sorry for the cats," says Miyazaki, referring to the experiment in which they placed cats on either end of a mosquito cage. "So, we did the experiment with our own arms. One had nepetalactol, the other only had the solvent ethanol. We put both arms in the cage. And we found that nepetalactol also protects our skin from mosquito bites."

Following that result, Iwate University and Nagoya University filed a patent, and they are looking for a pharmaceutical company that can turn the cat-discovered chemicals into an effective mosquito repellent for humans. If successful, Miyazaki and Nishikawa think that a potential human repellent will consist of multiple iridoids rather than nepetalactol alone.

"Our research shows that the mix of chemicals is better," says Miyazaki. "The other chemicals enhance the potent activity of nepetalactol, and that means that smaller amounts can be used."

As Uenoyama and Miyazaki have so aptly shown, animals still hold many medicinal secrets. By studying their behaviors in detail, we can find new chemicals that we can use to protect ourselves from infectious disease. What is so exciting about their work is that it provides a new and creative framework for the pharmaceutical industry to develop drugs. While animals have long taught humans how to develop herbal treatments, Uenoyama and Miyazaki have demonstrated that animals can do much more. They can point us at the exact chemicals that are responsible for the medicinal effects of the natural products they use.

The importance of this cannot be overstated. The pharmaceutical industry is built on the premise that drugs and insect repellents should be based on purified chemicals that can be delivered in carefully dosed and controlled pills or liquids—as opposed to dried herbs or herbal concoctions that characterize traditional medicine. By showing that animals can deliver those chemicals to us, Uenoyama and Miyazaki have started what I expect to be a long list of animal-discovered drugs.

14

Plants and Pollinators

Humans are latecomers to this earth. The first animals evolved more than 500 million years ago,[1] but our own species, *Homo sapiens*, has only been here for some two hundred thousand years. Millions of animal species were practicing medicine before we even evolved. Yet, despite our short time on this planet, and our late rise to the practice of medicine, we are currently destroying other species at an alarming rate, taking away the medicines and medicinal wisdom of thousands of animals. And that is not only bad for our fellow creatures. It also means we are destroying our own future pharmacy.

Species come and species go. Since the origin of life around 3.8 billion years ago, millions of species have evolved, and millions have gone extinct. When conditions change, new enemies evolve, or accidents happen, some species do not keep up and go extinct. That is normal, and scientists refer to the rate at which this happens as the background extinction rate. Sometimes, however, the earth experiences major events during which many more species perish. Our planet has experienced five mass extinctions, defined as events in which at least 60 percent of the species go extinct in less than 1 million years. Perhaps the most famous is the end-Cretaceous extinction about 65 million years ago, when most dinosaurs—along with more than 60 percent of other multicellular organisms—were wiped out by an asteroid that was roughly the size of Mount Everest and hit the earth with the impact of several million nuclear bombs. It was not the biggest extinction, however. That honor goes to the end of the Permian, 252 million years ago. No less than nine out of ten species disappeared at that time, most likely because of outpourings of molten rock that resulted in acid rain, widespread coal fires, mercury-filled air, and a lack of oxygen in the water.

Unfortunately, we are currently experiencing a sixth mass extinction, and this one is caused entirely by humans. According to the International Union for Conservation of Nature (IUCN), 338 vertebrates (mammals, birds, reptiles, amphibians, and fish) went extinct between 1500 and 2015, most of them in the last one hundred years. Another 279 species have either become "extinct in the wild" or "possibly extinct." Estimating extinction rates is difficult, but even the most conservative analysis of vertebrates shows that current extinction rates are eight to one hundred times higher than the background extinction rate.[2] This means that the vertebrate species we lost in the last century would normally have taken eight hundred to ten thousand years to have gone extinct.

And extinction is but one problem: because of overexploitation, habitat destruction, climate change, and the spread of invasive species and diseases, most species are now less abundant than they were before the industrial revolution.[3] The World Wildlife Fund's (WWF) *Living Planet Report 2020* states that humans have already altered 75 percent of the earth's ice-free land surface, and that population sizes of mammals, birds, reptiles, amphibians, and fish have fallen by 68 percent since 1970.[4]

According to another WWF report, over 43 million hectares of forest were lost in tropical and subtropical regions in Latin America, Sub-Saharan Africa, Southeast Asia, and Oceania between 2004 and 2017, roughly corresponding to the size of Morocco.[5] On top of that, much of the remaining forest is becoming degraded. Conversion of rainforest to oil palm plantations, combined with forest degradation and hunting, led to the loss of one hundred thousand Bornean orangutans between 1999 and 2015.[6] Given these massive declines in animal populations, researchers have coined the word "defaunation," an analogy with the well-known term "deforestation."

Vertebrates are much better studied than invertebrates, but our boneless relatives are not faring much better. For the limited invertebrates assessed by the IUCN, 40 percent are threatened. Butterfly and moth populations have declined by 35 percent globally, and invertebrate populations overall have declined by 45 percent since 1970.[7] The decline of insects is especially concerning, as these animals are crucial for the functioning of ecosystems. They are prey to other animals; they pollinate flowers and crops; they consume waste; and they recycle nutrients. They are also key for human survival, providing products such as honey, wax, dyes, and silk, and being used as biocontrol in agriculture. They provide a value of at least $70 billion (2020 dollars) to humans and agriculture in the United States on a yearly basis.[8]

This loss of species is not just sad. It is also detrimental to medicine. Having less access to medicinal plants jeopardizes the knowledge that underlies traditional medicine. In her book *Seeds of Hope*, Jane Goodall describes a traditional healer of the Waha tribe in Tanzania, who had to travel 160 miles to find a plant that used to grow close to home.[9] She describes this event not as an exceptional occurrence, but as a test case of what many healers are experiencing. Plants that used to be readily available are becoming scarce and challenging to find.

As if species loss is not bad enough, many Indigenous languages are also going extinct. Scientists estimate that we may lose more than a third of the roughly seven thousand documented languages by the end of this century, partly because of increasing connectivity between remote locations and larger towns.[10] This is especially troublesome because most traditional medicinal knowledge is passed orally between generations: an analysis published in 2021 showed that three-quarters of more than twelve thousand existing plant medicinal uses are known in only one unique language.[11] When that language disappears, so does the medicinal treatment. This is why Michael A. Huffman has embarked on preparing a three-way dictionary for the WaTongwe to document their medicinal knowledge in the local tongue, as well as in Swahili and English. It is also why ethnobotanists, such as Cassandra Quave, are so keen to document the medicinal practices of Indigenous people around the globe—before it is too late.

The biodiversity crisis is equally damaging to the natural medicine cabinet for animals. Loss of forests and habitats that are fragmented make it harder for wild animals to find their drugs. And we are not just limiting the abilities of wild species to medicate, but also those of companion and domesticated animals. Whether through urban expansion, agriculture, or domestication, we are forcing animals into depauperate habitats where they are unable to access medicinal plants. This is as true for the dogs we keep in our homes as for the cattle we keep in industrialized stables, the bees we truck to monocultures of almond trees, and the gorillas we keep in zoos. Not only do we prevent these animals from roaming free so they can collect their own medicine, we often discourage their natural tendencies to medicate themselves. You may remember the beekeepers who do not like working with sticky resin, and dog owners who discourage their dogs from eating dirt and grass.

And here is a scary realization: it is not just medicines we are losing. We are losing doctors as well. With every animal we lose, we lose another medical expert and another opportunity to discover new drugs that we may want to develop for ourselves. Simply put: fewer animals means fewer doctors we can learn from.

POLLINATOR GARDENS

While I am truly worried about the state of our planet, I believe that the discoveries that animals use medicine, and the insights that we can gain from them, gives us yet another reason to protect nature. It gives us one more incentive to keep this wonderful world of medical knowledge intact for our children and grandchildren. The beauty is that while we may have different reasons to preserve the natural pharmacy and its animal doctors, they all serve the same outcome. For some of us, animal medication may simply reinforce our awe of nature and our deep-rooted belief that we need to protect nature for the sake of it. For others, the health and economic benefits for humans may be the primary driver for conservation. Understanding that the next life-saving—and money-making—drug is out there, and that we can find it by studying the species we decide to preserve, can be a major incentive for conservation.

The biodiversity and climate crises we are experiencing can be overwhelming, and we may feel helpless in solving them. True, many of the necessary actions need to be taken by governments, including curbing deforestation, scaling back the use of fossil fuels, and developing clean transportation. But there are things we can do ourselves. We can support conservation organizations, invest in solar panels, and make changes to our diet (even replacing some beef meals with pork or chicken can go a long way in reducing carbon emissions and climate change).

Another thing we can do is to get rid of the idea that humans are different from animals and somehow superior. I grew up in a Western culture that views humans as special, important, and better than other species. Many people in my culture believe that God created this world for us, and that all the other species are here to serve us. When I first learned about evolution, the science was presented in a way that suggests that humans are the pinnacle of evolution, with all other species somehow being less evolved than us. I do not agree with either of these views. When I took religion classes in primary school, my teacher taught me that the Christian God she believed in created us to be the stewards of the natural world. When I studied biology, I learned that humans are the product of an evolutionary process that started 3.8 billion years ago. We are part of a massive tree of life, having evolved just as much as all the other creatures we share a world with.

Viewing ourselves as a member of a large family of animals that can share knowledge and learn from each other has medical benefits for both animals and humans. As cardiologist Barbara Natterson-Horowitz and science writer Kathryn Bowers describe in their book *Zoobiquity*, many of the diseases and health problems that afflict humans are equally found in animals.[12] By connecting to each other's knowledge, medical and veterinary doctors could make enormous progress in increasing the health of humans and animals alike. Viewing ourselves as a member of a large animal family can also increase our respect for animals and give us the motivation to protect them.[13] As

Huffman states in one of his papers: "As long as people are curious about nature and believe it has something to teach us, we have a chance at saving it along with ourselves."[14]

To mitigate the losses of nature we have caused, we can recreate diversity. Different countries have now developed agricultural schemes that help increase floral diversity and consequently ecosystem services, such as pollination. This includes the UK agri-environment scheme and US conservation reserve programs. One of the plants included in the UK program is sainfoin (*Onobrychis viciifolia*), which has caffeine in its nectar. Researchers have shown that this nectar can work medicinally by reducing both the parasite load of *Nosema bombi* in individual bumblebees (*Bombus terrestris*) and lower the parasite prevalence in colonies.[15] This shows that by using ecological recommendations, we can provide medication to pollinators.

The great thing is that we do not need to rely on bigger organizations and governments to recreate nature. We can also create gardens ourselves. When I take my miniature poodle, Tukkie, for walks in my neighborhood, I pass by lawn after lawn. It always strikes me that the average homeowner in my community is obsessed with having monocultures of a single species of grass to create a perfect, green, flat space to surround their homes. These lawns are truly biodiversity deserts, with not much life other than the grass of choice.

Scientists have estimated that 164,000 square kilometers of land are occupied by turf grasses in the United States alone. To put that in perspective: this is more than the area covered by corn in the United States, and four times the size of the Netherlands.[16] On a yearly basis, the water needed to keep these lawns green could fill 5 million Olympic swimming pools. Most grasses used are not native. They need herbicides and fertilizers to do well. Nitrogen fertilizers cause massive carbon emissions, and nitrogen runoff, which results in toxic algal blooms in freshwater and coastal areas. Gas-powered mowers produce 5 percent of the nation's air pollution.

But here's the good news: even by replacing just 10 percent of our lawns with diverse plant gardens, we can boost biodiversity, maintain insects, and provide medicinal plants for our pets.[17] We may even use these gardens to grow medicinal plants for human use, which would help curb the overharvesting of some medicinal plants in the wild, and recreate the natural pharmacies that traditional healers rely on.[18] These gardens reduce the need for watering, and the insects that they attract can take over the role of pesticides. By adding the right species, we can scare away mosquitoes (try catnip!). Meanwhile, our dogs and cats may find antiparasitic and calming plants in these gardens. Diverse flower gardens support native pollinators and honey bees, and preserve our crop production.

Because of all these benefits, local and national governments are increasingly developing initiatives for individual households to create native habitats. In Germany, the Thousands of Gardens—Thousands of Species project aims to create oases of biodiversity, including gardens, balconies, and open spaces with a goal to curb insect declines.[19] The project is funded in part by Germany's federal government and partners with seed companies, nurseries, and garden centers to provide seed packages to participants. In Minnesota, the Lawns to Legumes program provides grants to homeowners to develop natural gardens. In my home country of the Netherlands, an organization called The Pollinators provides free bags of seed mixes to create insect gardens.

In Georgia, my new adopted home, I am on the board of directors of the Rosalynn Carter Butterfly Trail, a nonprofit organization that aims to expand pollinator habitats. Initiated

FIGURE 14.1. Even converting a small part of a lawn can go a long way in providing medicine to pollinators and pets. Photo by Dan4Earth / Shutterstock.

by former first lady Rosalynn Carter, the trail was established in 2013 and grew to almost two thousand gardens in the first ten years of its existence. As an organization, we provide guidance on what plants to use and how to maintain the gardens. We encourage people to plant native milkweeds to support monarch butterflies. My own lab provides milkweeds and other plants for participants in the metro-Atlanta area. It may seem like a small thing to create a garden, but if enough people do it, we can recreate a lot of much-needed nature.

One common theme in this book has been that animals need choices. They do not just need shelter and food. They need medicine. And to get that, they need access to a diversity of plants and other natural products. Preserving nature is the best way to maintain their choices, and so is providing diverse gardens to pets, zoo animals, and our neighborhood insects.

So, as you are building your garden, I invite you to take a moment to witness the spectacles of the natural world that will unfold there. Live in the moment and witness that ant, bee, or butterfly that visits your garden. Watch your cat or dog frolic in the flowers. And as you are watching, ask yourself: What is the animal doing? Is it eating? Is it drinking? Is it finding shelter?

Or, maybe: Is it collecting medicine?

ACKNOWLEDGMENTS

This book would not have been possible without the many colleagues, postdocs, and students who have worked with me on monarchs, parasites, and medication over the years, including Tarik Acevedo, Kandis Adams, Aamina Ahmad, Wajd Alaidrous, Tiffany Alcaide, Samuel Alizon, Ahmed Aljohani, Emmanuel Arega, Tolu Babalola, Paola Barriga, Becky Bartel, Chris Catano, Joselyne Chavez, Jean Chi, Allen Chiang, Andy Davis, Leslie Decker, Gabe DuBose, Marissa Duckett, Ali Ebada, Tyler Faits, Lydia Fuller-Hall, Nicole Gerardo, Laura Gold, Camden Gowler, Erica Harris, Amy Hastings, Kevin Hoang, Miles Hollimon, Mackenzie Hoogshagen, Junjian Huang, Maggie Kelavkar, Kieran Kelly, Mitchell Kendzel, Krish Khurana, Yaw Kumi-Ansu, Kristoffer Leon, Helen Li, Hui Li, James Li, Carlos Lopez Fernandez de Castillejo, Justine Lyons, Ania Majewska, Andrew Mongue, Karen Oberhauser, Lindsay Oliver, Tristan Olpin, Andrew Pahnke, Amy Pedersen, Rayshaun Pettit, Amanda Pierce, Karl Protil III, Rachel Rarick, Amanda Rawstern, Dara Satterfield, Sarah Sonny, Eleanore Sternberg, Elizabeth Sun, Rohini Swamy, Wen-Hao Tan, Leiling Tao, Michelle Tsai, Milan Udawatta, Scott Villa, James Walters, Rebecca Wang, Amanda Williams, Andy Yates, and Ella Zhao.

I also want to thank the staff and volunteers at the St. Marks National Wildlife Refuge, who have invited me year after year to join them for the annual St. Marks Monarch Butterfly

Festival, including David Cook, Lori Nicholson, Terry Peacock, and Robin Will. The late Ms. Rosalynn Carter, and Annette Wise deserve special thanks for inviting me to join the board of the Rosalynn Carter Butterfly Trail, which has shaped much of my thinking on recreating natural habitats.

Special thanks go to Mark Hunter, who planted the first seeds in my mind that monarchs may use milkweeds as medicine, and whose ecological chemistry expertise has been instrumental in our understanding of monarch medication. I also want to give special thanks to Sonia Altizer, who invited me to join her lab to start working on monarchs, and Thierry Lefèvre, with whom I carried out the first experiments that showed that monarch butterflies can medicate their offspring.

I am especially grateful to Michael A. Huffman. Pioneering the field of animal self-medication, Mike has been an inspiration to many. As I was writing this book, he became a personal mentor to me. We corresponded by email and met on Zoom many times, not only to discuss his own work on chimpanzees and other animals but also to bounce around all sorts of ideas on animal medication. He connected me with many other people whom I interviewed and featured in this book. In March 2023, I had the privilege to visit his lab at Kyoto University, where he generously hosted me for ten days so I could pick his brain, use his personal library, benefit from his review of all my chapter drafts, and drink different varieties of medicinal green tea.

Many people gracefully met with me in person or on Zoom to discuss their work on animal medication and then provided comments on chapter drafts: Michel Chapuisat, Philippe Christe, Maria Fernanda De la Fuenta, Carly Hillier, Caroline Ingraham, Constantino Macías Garcia, Ulrich Maloueki,

Masao Miyazaki, Fred Provenza, Cassandra Quave, Mike Simone-Finstrom, Mike Singer, Marla Spivak, Monserrat Suárez Rodríguez, Reiko Uenoyama, and Juan Villalba.

Other people were instrumental in my getting started on this project. Anurag Agrawal and Menno Schilthuizen gave me helpful advice on developing a book proposal. The late Frans de Waal, and Catherine Marin read early drafts of chapters and helped me find my voice as I started writing this book. Mans Kuipers helped me develop my writing skills when I was an aspiring science writer many years ago, and Renee Braams first made me believe that writing a book would be a realistic option.

Special thanks to my editors Alison Kalett and Hallie Schaeffer. I first met Alison when I gave a research seminar on monarch medication at Princeton University in 2014. Sitting at a picnic table in the March sun, Alison asked me if I would consider writing a book. I told her the time was not right, but her question kept lingering. When I contacted her seven years later, she promptly responded with great enthusiasm and helpful guidance in writing my book proposal. In addition to their enthusiasm and excitement about this project, Alison and Hallie provided many rounds of helpful comments on chapter and manuscript drafts, helped me shape many different stories and ideas into a coherent book, and guided me in making the manuscript more accessible for readers who are not experts on the science of animal medication.

Lots of other people also provided me with valuable feedback on chapter drafts, including Jason Chen, Ali Ebada, Erik Edwards, Nicole Gerardo, Mackenzie Hoogshagen, Anthony Junker, Kieran Kelly, David Kennedy, Sandy Lin, Ania Majewska, and Scott Villa. Special thanks go to LM Bradley, Mitchell

Kendzel, Michelle McCauley, and Sandra Mendiola, who read multiple drafts of my full manuscript to give me feedback on structure, flow, and organization. Thanks also to Ludo Hellemans, who pushed me to be more careful in my historical assessments of Darwin and Wallace. I am also indebted to three anonymous reviewers, who provided constructive and enthusiastic feedback that helped me improve the manuscript.

No book can be a complete overview of a scientific field, and this book is no exception. As I developed the narrative, I had to make choices. I needed to feature some stories and leave others out. I ended up interviewing some scientists, while excluding others. As I was trimming the long lists of examples, I had to cut many interesting stories on capuchins, millipedes, anting birds, finches, drug discovery, ethnobotany, and history. To the many scientists whose research I ended up not including, I apologize.

I have benefited from a wonderful support team over the years. Family members, including Jakob, Ella, and Daphne, were there for me in times of need. And as I was diving into the literature on animal medication, they patiently spent many dinners listening to my new discoveries, my ideas for this book, and my new obsession with bears and how they shaped human medicine, culture, and even language (I highly recommend Wolf Storl's book *Bear: Myth, Animal, Icon*!).

Many other people supported me as I was juggling the writing of this book with family life, a university professor job, the COVID-19 pandemic, and societal issues. Special thanks go to Brooke Lewallen, Steven L'Hernault, Miguel Reyes, and Jessica Schmoll.

Finally, I want to thank Tukkie, in my opinion the cutest puppy in the world. Tukkie joined my family shortly before

I embarked on this book project, and his eating of grass and dirt inspired many of my investigations. He also spent many hours sitting in my lap while I was reading books and papers, talking with people on Zoom, and writing and editing chapters. I am not sure I would have finished this project without his unconditional love.

NOTES

1. Birds, Bees, and Butterflies

1. Brower, L. P. Understanding and misunderstanding the migration of the monarch butterfly (Nymphalidae) in North America: 1857–1995. *Journal of the Lepidopterists' Society* **49**, 304–385 (1995).

2. McLaughlin, R. E. & Myers, J. *Ophryocystis elektroscirrha* sp. n., a neogregarine pathogen of monarch butterfly *Danaus plexippus* (L.) and the Florida queen butterfly *D. gilippus berenice* Cramer. *Journal of Protozoology* **17**, 300–305 (1970).

3. Bartel, R. A., Oberhauser, K. S., de Roode, J. C. & Altizer, S. Monarch butterfly migration and parasite transmission in eastern North America. *Ecology* **92**, 342–351 (2011).

4. Agrawal, A. A. *Monarchs and Milkweed* (Princeton University Press, 2017).

5. Malcolm, S. B. & Brower, L. P. Evolutionary and ecological implications of cardenolide sequestration in the monarch butterfly. *Experientia* **45**, 284–295 (1989).

6. Brower, L. P., Ryerson, W. N., Coppinger, L. & Glazier, S. C. Ecological chemistry and the palatability spectrum. *Science* **161**, 1349–1351 (1968).

7. Cory, J. S. & Hoover, K. Plant-mediated effects in insect-pathogen interactions. *Trends in Ecology & Evolution* **21**, 278–286 (2006); Keating, S. T., Hunter, M. D. & Schultz, J. C. Leaf phenolic inhibition of gypsy moth nuclear polyhedrosis virus—role of polyhedral inclusion body aggregation. *Journal of Chemical Ecology* **16**, 1445–1457 (1990).

8. de Roode, J. C., Pedersen, A. B., Hunter, M. D. & Altizer, S. Host plant species affects virulence in monarch butterfly parasites. *Journal of Animal Ecology* **77**, 120–126 (2008).

9. Lefèvre, T., Oliver, L., Hunter, M. D. & de Roode, J. C. Evidence for transgenerational medication in nature. *Ecology Letters* **13**, 1485–1493 (2010).

10. de Roode, J. C., Chi, J., Rarick, R. M. & Altizer, S. Strength in numbers: high parasite burdens increase transmission of a protozoan parasite of monarch butterflies (*Danaus plexippus*). *Oecologia* **161**, 67–75 (2009).

11. MacIntosh, A. J. & Huffman, M. A. Toward understanding the role of diet in host–parasite interactions: The case for Japanese Macaques. In: *The Japanese Macaques* (ed. F. Nakagawa, M. Nakamichi, & H. Sugiura) pp. 323–344 (Springer, 2010).

12. Phillips-Conroy, J. E. Baboons, diet and disease: food plant selection and schistosomiasis. In: *Current Perspectives in Primate Social Dynamics* (ed. D. M. Taub & F. A. King) pp. 287–304 (Van Nostrand Reinhold, 1986).

13. Rodriguez, E. & Wrangham, R. Zoopharmacognosy: the use of medicinal plants by animals. *Recent Advances in Phytochemistry* **27**, 89–105 (1993).

14. Mascaro, A., Southern, L. M., Deschner, T. & Pika, S. Application of insects to wounds of self and others by chimpanzees in the wild. *Current Biology* **32**, R112–R113 (2022).

15. Laumer, I. B. et al. Active self-treatment of a facial wound with a biologically active plant by a male Sumatran Orangutan. *Scientific Reports* **14**, 8932, https://doi.org/10.1038/s41598-024-58988-7 (2024).

16. Morrogh-Bernard, H. et al. Self-medication by orang-utans (*Pongo pygmaeus*) using bioactive properties of *Dracaena cantleyi*. *Scientific Reports* **7**, 16653 (2017).

17. Carrai, V., Borgognini-Tarli, S. M., Huffman, M. A. & Bardi, M. Increase in tannin consumption by sifaka (*Propithecus verreauxi verreauxi*) females during the birth season: a case for self-medication in prosimians? *Primates* **44**, 61–66 (2003).

18. *50 drunken elephants ransack village in India, drink 130 gallons of moonshine.* Fox8 News (website), November 7, https://myfox8.com/news/50-drunken-elephants-ransack-village-in-india-drink-130-gallons-of-moonshine/ (2012).

19. Amato, K. R. et al. Fermented food consumption in wild nonhuman primates and its ecological drivers. *American Journal of Physical Anthropology* **175**, 513–530 (2021).

20. Engel, C. *Wild Health: Lessons in Natural Wellness from the Animal Kingdom* (Houghton Mifflin Harcourt, 2002).

21. Hughes, D. P., Brodeur, J. & Thomas, F. *Host Manipulation by Parasites* (Oxford University Press, 2012).

22. Kirsch, D. R. & Ogas, O. *The Drug Hunters* (Arcade Publishing, 2018).

23. Newman, D. J. & Cragg, G. M. Natural products as sources of new drugs over the nearly four decades from 01/1981 to 09/2019. *Journal of Natural Products* **83**, 770–803 (2020).

2. Chimp Chausiku

1. Osler, W. *Aequanimitas: Teaching and Thinking* (H. K. Lewis & Co., 1914).

2. Chapman, C. A. & Huffman, M. A. Why do we want to think humans are different? *Animal Sentience* **3**, 163 (2018).

3. de Waal, F. *Are We Smart Enough to Know How Smart Animals Are?* (Norton, 2017).

4. Zuk, M. *Dancing Cockatoos and the Dead Man Test: How Behavior Evolves and Why It Matters* (W. W. Norton, 2022).

5. Kawai, M. Newly-acquired pre-cultural behavior of the natural troop of Japanese monkeys on Koshima Islet. *Primates* **6**, 1–30 (1965).

6. de Waal, F. *The Ape and the Sushi Master* (Basic Books, 2001).

7. Storl, W. D. *Bear: Myth, Animal, Icon* (North Atlantic Books, 2018); Rockwell, D. *Giving Voice to Bear: North American Indian Myths, Rituals, and Images of the Bear* (Roberts Reinhart, 1991).

8. Storl, *Bear*; Rockwell, *Giving Voice to Bear*; Lake-Thom, B. *Spirits of the Earth: A Guide to Native American Nature Symbols, Stories, and Ceremonies* (Plume, 1997); Lame Deer, J. F. & Erdoes, R. *Lame Deer, Seeker of Visions* (Simon & Schuster, 2009); Densmore, F. *Uses of Plants by the Chippewa Indians* (US Government Printing Office, 1928).

9. Densmore, *Uses of Plants*.

10. Densmore, *Uses of Plants*.

11. Lame Deer & Erdoes. *Lame Deer Seeker of Visions*.

12. Storl, *Bear*.

13. Clayton, D. H. & Wolfe, N. D. The adaptive significance of self-medication. *Trends in Ecology & Evolution* **8**, 60–63 (1993); Myhal, N. *Ethnobotany of Oshá (Ligusticum porteri) and Policy of Medicinal Plant Harvest on United States Forest Service Lands* (master's thesis, University of Kansas, 2017).

14. Myhal, *Ethnobotany of Oshá*.

15. Hellgren, E. C. Physiology of hibernation in bears. *Ursus* **10**, 467–477 (1998).

16. Jeffreys, D. *Aspirin: The Remarkable Story of a Wonder Drug*, p. 10 (Bloomsbury, 2004).

17. Conversations between Michael A. Huffman and the author. By video call, September and October 2021, and in person in Japan, March 2023.

18. Huffman, M. Learning to become a monkey. In: *Primate Ethnographies* (ed. K. Strier) pp. 57–68 (Pearson Education, 2014).

19. de Waal, *Ape and Sushi Master*.

20. de Waal, *Ape and Sushi Master*; Asquith, P. J. Provisioning and the study of free-ranging primates: history, effects, and prospects. *Yearbook of Physical Anthropology* **32**, 129–158 (1989); Kitahara-Frisch, J. Culture and primatology: East and West. In: *The Monkeys of Arashiyama* (ed. L. M. Fedigan & P. J. Asquith) pp. 74–80 (State University of New York Press, 1991).

21. Huffman, M. A. The lessons of history and tradition: on becoming a monkey and other insights gained as a primatologist in Japan. *Primates* **64**, 5–8 (2023).

22. Ohigashi, H. et al. Bitter principle and a related steroid glucoside from *Vernonia amygdalina*, a possible medicinal plant for wild chimpanzees. *Agricultural and Biological Chemistry* **55**, 1201–1203 (1991); Jisaka, M. et al. Bitter steroid glucosides, vernoniosides A1, A2, and A3, and related B1 from a possible medicinal plant, *Vernonia amygdalina*, used by wild chimpanzees. *Tetrahedron* **48**, 625–632 (1992); Jisaka, M., Ohigashi, H., Takegawa, K. & Koshimizu, K. Antitumoral and antimicrobial activities of bitter sesquiterpene lactones of *Vernonia amygdalina*, a possible medicinal plant used by wild chimpanzees. *Bioscience, Biotechnology, and Biochemistry* **57**, 833–834 (1993).

23. Huffman, M. A., Gotoh, S., Izutsu, D., Koshimizu, K. & Kalunde, M. S. Further obervations on the use of the medicinal plant, *Vernonia amygdalina* (Del), by a wild chimpanzee, its possible effect on parasite load, and its phytochemistry. *African Study Monographs* **14**, 227–240 (1993).

24. Huffman, M. A., Gotoh, S., Turner, L. A., Hamai, M. & Yoshida, K. Seasonal trends in intestinal nematode infection and medicinal plant use among chimpanzees in the Mahale Mountains, Tanzania. *Primates* **38**, 111–125 (1997).

25. Ohigashi et al., Bitter principle; Huffman et al., Further obervations on the use of *Vernonia amygdalina*; Huffman, M. A. Chimpanzee self-medication: a historical perspective of the key findings. In: *Mahale Chimpanzees: 50 Years of Research* (ed. M. Nakamura, K. Hosaka, N. Itoh, & K. Zamma) pp. 340–353 (Cambridge University Press, 2015); Huffman, M. A. & Seifu, M. Observations on the illness and consumption of a possibly medicinal plant *Vernonia amygdalina* (Del.), by a wild chimpanzee in the Mahale Mountains National Park, Tanzania. *Primates* **30**, 51–63 (1989); Huffman, M. & Caton, J. Self-induced increase of gut motility and the control of parasitic infections in wild chimpanzees. *International Journal of Primatology* **22**, 329–346 (2001); Koshimizu, K., Ohigashi, H. & Huffman, M. A. Use of *Vernonia amygdalina* by wild chimpanzee: possible roles of its bitter and related constituents. *Physiology & Behavior* **56**, 1209–1216 (1994).

26. "TEDxOsaka: Michael Huffman—Animal Self-medication," TEDx Talks, July 10, 2012, https://www.youtube.com/watch?v=WNn7b5VHowM&t=689s.

27. Huffman, Chimpanzee self-medication.

28. Goodall, J. *Seeds of Hope: Wisdom and Wonder from the World of Plants* (Grand Central Publishing, 2014).

29. Wrangham, R. W. & Nishida, T. *Aspilia* spp. leaves: a puzzle in the feeding behavior of wild chimpanzees. *Primates* **24**, 276–282 (1983).

30. Rodriguez, E. et al. Thiarubrine A, a bioactive constituent of *Aspilia* (Asteraceae) consumed by wild chimpanzees. *Experientia* **41**, 419–420 (1985).

31. Rodriguez, E. & Wrangham, R. Zoopharmacognosy: the use of medicinal plants by animals. *Recent Advances in Phytochemistry* **27**, 89–105 (1993).

32. Huffman, Chimpanzee self-medication; Huffman, M. A. et al. Leaf-swallowing by chimpanzees: a behavioral adaptation for the control of strongyle nematode infections. *International Journal of Primatology* **17**, 475–503 (1996).

33. Dupain, J. et al. New evidence for leaf swallowing and *Oesophagostomum* infection in bonobos (*Pan paniscus*). *International Journal of Primatology* **23**, 1053–1062 (2002).

34. Huffman, M. A. An ape's perspective on the origins of medicinal plant use in humans. In: *Wild Harvest: Plants in the Hominin and Pre-Agrarian Human Worlds* (ed. K. Hardy & L. Kubiak-Martens) pp. 55–70 (Oxbow Books, 2016).

35. Huffman et al., Seasonal trends in intestinal nematode infection.

36. Wrangham, R. W. Relationship of chimpanzee leaf-swallowing to a tapeworm infection. *American Journal of Primatology* **37**, 297–303 (1995).

37. Huffman & Caton, Self-induced increase of gut motility.

38. Huffman, An ape's perspective.

39. Barelli, C. & Huffman, M. A. Leaf swallowing and parasite expulsion in Khao Yai white-handed gibbons (*Hylobates lar*), the first report in an Asian ape species. *American Journal of Primatology* **79**, e22610 (2017).

3. Parasites and Pathogens

1. Windsor, D. A. Most of the species on Earth are parasites. *International Journal for Parasitology* **28**, 1939–1941 (1998).

2. Gering, E. et al. *Toxoplasma gondii* infections are associated with costly boldness toward felids in a wild host. *Nature Communications* **12**, 3842 (2021); Brandell, E. E. et al. Infectious disease in Yellowstone's wolves. In: *Yellowstone Wolves: Science Discovery in the World's First National Park* (ed. D. Smith, D. Stahler & D. MacNulty) pp. 121–133 (University of Chicago Press, 2020).

3. Gryseels, B., Polman, K., Clerinx, J. & Kestens, L. Human schistosomiasis. *The Lancet* **368**, 1106–1118 (2006).

4. Cory, J. S. & Hoover, K. Plant-mediated effects in insect-pathogen interactions. *Trends in Ecology & Evolution* **21**, 278–286 (2006).

5. Adelman, J. S., Moyers, S. C., Farine, D. R. & Hawley, D. M. Feeder use predicts both acquisition and transmission of a contagious pathogen in a North American songbird. *Proceedings of the Royal Society of London Series B: Biological Sciences* **282**, 20151429 (2015).

6. Chaffin, T. *Odyssey: Young Charles Darwin, the Beagle, and the Voyage that Changed the World* (Pegasus: 2022).

7. Wallace, A. R. On the tendency of varieties to depart indefinitely from the original type. *Journal of the Proceedings of The Linnean Society London, Zoology* **3**,

45–62 (1858); Darwin, C. Extract from an unpublished work on species, by C. Darwin, Esq., consisting of a portion of a chapter entitled, "On the variation of organic beings in a state of nature; on the natural means of selection; on the comparison of domestic races and true species."—Abstract of a letter from C. Darwin, Esq., to Prof. Asa Gray, Boston, U.S., dated Down, September 5th, 1857. *Journal of the Proceedings of The Linnean Society London, Zoology* **3**, 45–62 (1858).

8. Dobzhansky, T. Nothing in biology makes sense except in the light of evolution. *American Biology Teacher* **35**, 125–129 (1973).

9. Donihue, C. M. et al. Hurricane effects on Neotropical lizards span geographic and phylogenetic scales. *Proceedings of the National Academy of Sciences* **117**, 10429–10434 (2020).

10. Jablonski, N. G. & Chaplin, G. Human skin pigmentation as an adaptation to UV radiation. *Proceedings of the National Academy of Sciences* **107**, 8962–8968 (2010).

11. Taubenberger, J. K. & Kash, J. C. Influenza virus evolution, host adaptation, and pandemic formation. *Cell Host & Microbe* **7**, 440–451 (2010).

12. Parker, B. J., Spragg, C. J., Altincicek, B. & Gerardo, N. M. Symbiont-mediated protection against fungal pathogens in pea aphids: a role for pathogen specificity? *Applied and Environmental Microbiology* **79**, 2455–2458 (2013); Scarborough, C. L., Ferrari, J. & Godfray, H. Aphid protected from pathogen by endosymbiont. *Science* **310**, 1781–1781 (2005).

13. Harris, R. N., James, T. Y., Lauer, A., Simon, M. A. & Patel, A. Amphibian pathogen *Batrachochytrium dendrobatidis* is inhibited by the cutaneous bacteria of amphibian species. *EcoHealth* **3**, 53–56 (2006).

14. Mitoh, S. & Yusa, Y. Extreme autotomy and whole-body regeneration in photosynthetic sea slugs. *Current Biology* **31**, R233–R234 (2021).

15. Curtis, V. A. Infection-avoidance behaviour in humans and other animals. *Trends in Immunology* **35**, 457–464 (2014).

16. Hart, B. L. Behavioral adaptations to pathogens and parasites: five strategies. *Neuroscience & Biobehavioral Reviews* **14**, 273–294 (1990).

17. Bush, S. E. & Clayton, D. H. Anti-parasite behaviour of birds. *Philosophical Transactions of the Royal Society B: Biological Sciences* **373**, 20170196 (2018).

18. Hart, Behavioral adaptations to pathogens and parasites.

19. Curtis, Infection-avoidance behaviour; Sarabian, C. et al. Disgust in animals and the application of disease avoidance to wildlife management and conservation. *Journal of Animal Ecology* **92**, 1489–1508 (2023).

20. Heinze, J. & Walter, B. Moribund ants leave their nests to die in social isolation. *Current Biology* **20**, 249–252 (2010).

21. Pusceddu, M. et al. Honey bees increase social distancing when facing the ectoparasite *Varroa destructor*. *Science Advances* **7**, eabj1398 (2021).

22. Curtis, Infection-avoidance behaviour; Stockmaier, S. et al. Infectious diseases and social distancing in nature. *Science* **371**, eabc8881 (2021).

23. Van Valen, L. A new evolutionary law. *Evolutionary Theory* **1**, 1–30 (1973); Brockhurst, M. A. & Koskella, B. Experimental coevolution of species interactions. *Trends in Ecology & Evolution* **28**, 367–375 (2013).

24. Naundrup, A. et al. Pathogenic fungus uses volatiles to entice male flies into fatal matings with infected female cadavers. *The ISME Journal* **16**, 2388–2397 (2022).

4. Beetles and Bulldogs

1. de Roode, J. C., Pedersen, A. B., Hunter, M. D. & Altizer, S. Host plant species affects virulence in monarch butterfly parasites. *Journal of Animal Ecology* **77**, 120–126 (2008).

2. Thomas, F. et al. Do hairworms (Nematomorpha) manipulate the water seeking behaviour of their terrestrial hosts? *Journal of Evolutionary Biology* **15**, 356–361 (2002).

3. Lefèvre, T., Oliver, L., Hunter, M. D. & de Roode, J. C. Evidence for transgenerational medication in nature. *Ecology Letters* **13**, 1485–1493 (2010).

4. Poli, C. H. et al. Self-selection of plant bioactive compounds by sheep in response to challenge infection with *Haemonchus contortus*. *Physiology & Behavior* **194**, 302–310 (2018).

5. Schneider, D. S. & Ayres, J. S. Two ways to survive infection: what resistance and tolerance can teach us about treating infectious diseases. *Nature Reviews: Immunology* **8**, 889–895 (2008).

6. Centers for Disease Control and Prevention. *Recommendations for the Use of Antibiotics for the Treatment of Cholera*, https://www.cdc.gov/cholera/treatment/antibiotic-treatment.html (2022).

7. Raubenheimer, D. & Simpson, S. J. Nutritional PharmEcology: doses, nutrients, toxins, and medicines. *Integrative and Comparative Biology* **49**, 329–337 (2009).

8. Krief, S., Hladik, C. M. & Haxaire, C. Ethnomedicinal and bioactive properties of plants ingested by wild chimpanzees in Uganda. *Journal of Ethnopharmacology* **101**, 1–15 (2005); Krief, S. et al. Bioactive properties of plant species ingested by chimpanzees (*Pan troglodytes schweinfurthii*) in the Kibale National Park, Uganda. *American Journal of Primatology* **68**, 51–71 (2006).

9. Huashuayo-Llamocca, R. & Heymann, E. W. Fur-rubbing with *Piper* leaves in the San Martín titi monkey, *Callicebus oenanthe*. *Primate Biology* **4**, 127–130 (2017); Baker, M. Fur rubbing: use of medicinal plants by capuchin monkeys (*Cebus capucinus*). *American Journal of Primatology* **38**, 263–270 (1996); Valderrama, X., Robinson, J. G., Attygalle, A. B. & Eisner, T. Seasonal anointment with millipedes in a wild primate:

a chemical defense against insects? *Journal of Chemical Ecology* **26**, 2781–2790 (2000); Birkinshaw, C. R. Use of millipedes by black lumurs to anoint their bodies. *Folia Primatologica* **70**, 170 (1999).

10. Hoogshagen, M. et al. Mixtures of milkweed cardenolides protect monarch butterflies against parasites. *Journal of Chemical Ecology*, https://doi.org/10.1007/s10886-023-01461-y (2023); Tao, L., Hoang, K. M., Hunter, M. D. & de Roode, J. C. Fitness costs of animal medication: anti-parasitic plant chemicals reduce fitness of monarch butterfly hosts. *Journal of Animal Ecology* **85**, 1246–1254 (2016); Agrawal, A. A. et al. Cardenolides, toxicity, and the costs of sequestration in the coevolutionary interaction between monarchs and milkweeds. *Proceedings of the National Academy of Sciences* **118**, e2024463118 (2021).

11. Morlock, G. E. et al. Evidence that Indo-Pacific bottlenose dolphins self-medicate with invertebrates in coral reefs. *iScience* **25**, 104271 (2022).

12. Valderrama et al., Seasonal anointment with millipedes; Weldon, P. J., Aldrich, J. R., Klun, J. A., Oliver, J. E. & Debboun, M. Benzoquinones from millipedes deter mosquitoes and elicit self-anointing in capuchin monkeys (*Cebus* spp.). *Naturwissenschaften* **90**, 301–304 (2003); Santos, E. R., Ferrari, S. F., Beltrão-Mendes, R. & Gutiérrez-Espeleta, G. A. Anointing with commercial insect repellent by free-ranging *Cebus capucinus* in Manuel Antonio National Park, Quepos, Costa Rica. *Primates* **60**, 559–563 (2019).

13. Baker, Fur rubbing; Conversation by video call between M. F. De la Fuente and the author, 12 October 2022.

14. Bravo, C., Bautista, L. M., Garcia-Paris, M., Blanco, G. & Alonso, J. C. Males of a strongly polygynous species consume more poisonous food than females. *PLOS ONE* **9**, e111057 (2014).

5. Birds and Butts

1. Rajasekar, R., Chattopadhyay, B. & Sripathi, K. Depositing masticated plant materials inside tent roosts in *Cynopterus sphinx* (Chiroptera: Pteropodidae) in Southern India. *Acta Chiropterologica* **8**, 269–274 (2006).

2. Hemmes, R. B., Alvarado, A. & Hart, B. L. Use of California bay foliage by wood rats for possible fumigation of nest-borne ectoparasites. *Behavioral Ecology* **13**, 381–385 (2002).

3. Samson, D. R., Muehlenbein, M. P. & Hunt, K. D. Do chimpanzees (*Pan troglodytes schweinfurthii*) exhibit sleep related behaviors that minimize exposure to parasitic arthropods? A preliminary report on the possible anti-vector function of chimpanzee sleeping platforms. *Primates* **54**, 73–80 (2013).

4. Mennerat, A. et al. Aromatic plants in nests of the blue tit *Cyanistes caeruleus* protect chicks from bacteria. *Oecologia* **161**, 849–855 (2009); Yang, C. et al. Sparrows

use a medicinal herb to defend against parasites and increase offspring condition. *Current Biology* **30**, R1411–R1412 (2020); Ontiveros, D., Caro, J. & Pleguezuelos, J. Green plant material versus ectoparasites in nests of Bonelli's eagle. *Journal of Zoology* **274**, 99–104 (2008).

5. Clark, L. & Mason, J. R. Use of nest material as insecticidal and anti-pathogenic agents by the European starling. *Oecologia* **67**, 169–176 (1985).

6. Clark, L. & Mason, J. R. Effect of biologically active plants used as nest material and the derived benefit to starling nestlings. *Oecologia* **77**, 174–180 (1988).

7. Conversation by video call between M. Suárez Rodríguez, C. Macías Garcia, and the author, 12 January 2022.

8. Suárez-Rodríguez, M., López-Rull, I. & Macías Garcia, C. Incorporation of cigarette butts into nests reduces ectoparasite load in urban birds: new ingredients for an old recipe? *Biology Letters* **9**, 20120931 (2012).

9. Suárez-Rodríguez, M. & Macías Garcia, C. An experimental demonstration that house finches add cigarette butts in response to ectoparasites. *Journal of Avian Biology* **48**, 1316–1321 (2017).

10. Suárez-Rodríguez, M. & Macías Garcia, C. There is no such a thing as a free cigarette: lining nests with discarded butts brings short-term benefits, but causes toxic damage. *Journal of Evolutionary Biology* **27**, 2719–2726 (2014).

11. Lans, C. & Turner, N. Organic parasite control for poultry and rabbits in British Columbia, Canada. *Journal of Ethnobiology and Ethnomedicine* **7**, 21 (2011).

12. Green, D. S., Tongue, A. D. & Boots, B. The ecological impacts of discarded cigarette butts. *Trends in Ecology & Evolution* **37**, 183–192 (2021).

13. Green, Tongue & Boots, Ecological impacts of discarded cigarette butts.

6. Ants and Aliens

1. De la Fuente, M. F., Souto, A., Albuquerque, U. P. & Schiel, N. Self-medication in nonhuman primates: A systematic evaluation of the possible function of the use of medicinal plants. *American Journal of Primatology*, e23438 (2022).

2. Mascaro, A., Southern, L. M., Deschner, T. & Pika, S. Application of insects to wounds of self and others by chimpanzees in the wild. *Current Biology* **32**, R112–R113 (2022).

3. Stork, N. E. How many species of insects and other terrestrial arthropods are there on Earth. *Annual Review of Entomology* **63**, 31–45 (2018).

4. Misof, B. et al. Phylogenomics resolves the timing and pattern of insect evolution. *Science* **346**, 763–767 (2014).

5. Pickrell, J. The making of mammals. *Nature* **574**, 468–472 (2019); Brusatte, S. L., O'Connor, J. K. & Jarvis, E. D. The origin and diversification of birds. *Current Biology* **25**, R888–R898 (2015).

6. Chadzopulu, A., Koukouliata, A., Theodosopoulou, E. & Adraniotis, J. Unique mastic resin from Chios. *Progress in Health Sciences* **1**, 131–135 (2011).

7. Christe, P., Oppliger, A., Bancala, F., Castella, G. & Chapuisat, M. Evidence for collective medication in ants. *Ecology Letters* **6**, 19–22 (2003).

8. Chapuisat, M., Oppliger, A., Magliano, P. & Christe, P. Wood ants use resin to protect themselves against pathogens. *Proceedings of the Royal Society of London Series B: Biological Sciences* **274**, 2013–2017 (2007).

9. Castella, G., Chapuisat, M. & Christe, P. Prophylaxis with resin in wood ants. *Animal Behaviour* **75**, 1591–1596 (2008).

10. Conversation by video call between P. Christe, M. Chapuisat, and the author, 18 January 2022.

11. Brütsch, T. & Chapuisat, M. Wood ants protect their brood with tree resin. *Animal Behaviour* **93**, 157–161 (2014).

12. Poyet, M. et al. The invasive pest *Drosophila suzukii* uses trans-generational medication to resist parasitoid attack. *Scientific Reports* **7**, 1–8 (2017).

13. Mokaya, H. O., Bargul, J. L., Irungu, J. W. & Lattorff, H.M.G. Bioactive constituents, in *vitro* radical scavenging and antibacterial activities of selected *Apis mellifera* honey from Kenya. *International Journal of Food Science and Technology* **55**, 1246–1254 (2020).

14. Gherman, B. I. et al. Pathogen-associated self-medication behavior in the honeybee *Apis mellifera*. *Behavioral Ecology and Sociobiology* **68**, 1777–1784 (2014).

15. Bos, N., Sundstrom, L., Fuchs, S. & Freitak, D. Ants medicate to fight disease. *Evolution* **69**, 2979–2984 (2015).

16. Koch, H. et al. Host and gut microbiome modulate the antiparasitic activity of nectar metabolites in a bumblebee pollinator. *Philosophical Transactions of the Royal Society B: Biological Sciences* **377**, 20210162 (2022).

17. Brütsch, T., Jaffuel, G., Vallat, A., Turlings, T. C. & Chapuisat, M. Wood ants produce a potent antimicrobial agent by applying formic acid on tree-collected resin. *Ecology and Evolution* **7**, 2249–2254 (2017).

18. Cory, J. S. & Hoover, K. Plant-mediated effects in insect-pathogen interactions. *Trends in Ecology & Evolution* **21**, 278–286 (2006).

19. Smilanich, A. M. & Muchoney, N. D. Host plant effects on the caterpillar immune response. In: *Caterpillars in the Middle* (ed. R. J. Marquis & S. Koptur) pp. 449–484 (Springer Nature Switzerland, 2022).

20. Felton, G. W. & Duffey, S. S. Inactivation of baculovirus by quinones formed in insect-damaged plant tissues. *Journal of Chemical Ecology* **16**, 1221–1236 (1990).

21. Salazar-Jaramillo, L. & Wertheim, B. Does *Drosophila sechellia* escape parasitoid attack by feeding on a toxic resource? *PeerJ* **9**, e10528 (2021).

22. Hoover, K., Washburn, J. O. & Volkman, L. E. Midgut-based resistance of *Heliothis virescens* to baculovirus infection mediated by phytochemicals in cotton. *Journal of Insect Physiology* **46**, 999–1007 (2000).

23. Del Campo, M. L., Halitschke, R., Short, S. M., Lazzaro, B. P. & Kessler, A. Dietary plant phenolic improves survival of bacterial infection in *Manduca sexta* caterpillars. *Entomologia Experimentalis et Applicata* **146**, 321–331 (2013).

24. Conversation by video call between M. S. Singer and the author, 5 January 2022.

25. Hartmann, T. et al. Acquired and partially *de novo* synthesized pyrrolizidine alkaloids in two polyphagous arctiids and the alkaloid profiles of their larval food-plants. *Journal of Chemical Ecology* **30**, 229–254 (2004).

26. Singer, M. S., Carriere, Y., Theuring, C. & Hartmann, T. Disentangling food quality from resistance against parasitoids: diet choice by a generalist caterpillar. *The American Naturalist* **164**, 423–429 (2004).

27. Singer, M. S., Mace, K. C. & Bernays, E. A. Self-medication as adaptive plasticity: increased ingestion of plant toxins by parasitized caterpillars. *PLOS ONE* **4**, e4796 (2009).

7. Poisons and Proteins

1. Kutschera, W. & Rom, W. Ötzi, the prehistoric Iceman. *Nuclear Instruments and Methods in Physics Research Section B: Beam Interactions with Materials and Atoms* **164**, 12–22 (2000).

2. Pleszczyńska, M. et al. *Fomitopsis betulina* (formerly *Piptoporus betulinus*): the Iceman's polypore fungus with modern biotechnological potential. *World Journal of Microbiology and Biotechnology* **33**, 1–12 (2017).

3. Capasso, L. 5300 years ago, the Ice Man used natural laxatives and antibiotics. *The Lancet* **352**, 1864 (1998).

4. Hardy, K. et al. Neanderthal medics? evidence for food, cooking, and medicinal plants entrapped in dental calculus. *Naturwissenschaften* **99**, 617–626 (2012).

5. Kirsch, D. R. & Ogas, O. *The Drug Hunters* (Arcade Publishing, 2018).

6. Kirsch & Ogas, *Drug Hunters*.

7. Whiteman, N. *Most Delicious Poison: The Story of Nature's Toxins—from Spices to Vices* (Little, Brown Spark, 2023).

8. Agosta, W. *Bombardier Beetles and Fever Trees: A Close-Up Look at Chemical Warfare and Signals in Animals and Plants* (Addison-Wesley, 1996).

9. Tu, Y. The discovery of artemisinin (qinghaosu) and gifts from Chinese medicine. *Nature Medicine* **17**, 1217–1220 (2011).

10. Nicolaou, K. C., Dai, W. M. & Guy, R. K. Chemistry and biology of taxol. *Angewandte Chemie International Edition in English* **33**, 15–44 (1994); Wall, M. E. &

Wani, M. C. Camptothecin and taxol: from discovery to clinic. *Journal of Ethnopharmacology* **51**, 239–254 (1996).

11. Williams, J. A., Day, M. & Heavner, J. E. Ziconotide: an update and review. *Expert Opinion on Pharmacotherapy* **9**, 1575–1583 (2008).

12. Parkes, D. G., Mace, K. F. & Trautmann, M. E. Discovery and development of exenatide: the first antidiabetic agent to leverage the multiple benefits of the incretin hormone, GLP-1. *Expert Opinion on Drug Discovery* **8**, 219–244 (2013).

13. Whiteman, *Most Delicious Poison.*

14. Agrawal, A. A. *Monarchs and Milkweed* (Princeton University Press, 2017).

15. Raupp, M. J. Effects of leaf toughness on mandibular wear of the leaf beetle, *Plagiodera versicolora. Ecological Entomology* **10**, 73–79 (1985).

16. Wari, D., Aboshi, T., Shinya, T. & Galis, I. Integrated view of plant metabolic defense with particular focus on chewing herbivores. *Journal of Integrative Plant Biology* **64**, 449–475 (2022).

17. Ehrlich, P. R. & Raven, P. H. Butterflies and plants. *Scientific American* **216**, 104–114 (1967).

18. Cooper, S. M. & Owen-Smith, N. Effects of plant spinescence on large mammalian herbivores. *Oecologia* **68**, 446–455 (1986).

19. White, C. & Eigenbrode, S. Effects of surface wax variation in *Pisum sativum* on herbivorous and entomophagous insects in the field. *Environmental Entomology* **29**, 773–780 (2000).

20. Kariyat, R. R., Smith, J. D., Stephenson, A. G., De Moraes, C. M. & Mescher, M. C. Non-glandular trichomes of *Solanum carolinense* deter feeding by *Manduca sexta* caterpillars and cause damage to the gut peritrophic matrix. *Proceedings of the Royal Society B: Biological Sciences* **284**, 20162323 (2017).

21. Chomel, M. et al. Plant secondary metabolites: a key driver of litter decomposition and soil nutrient cycling. *Journal of Ecology* **104**, 1527–1541 (2016).

22. Soltis, P. S. & Soltis, D. E. The origin and diversification of angiosperms. *American Journal of Botany* **91**, 1614–1626 (2004).

23. Hartmann, T. From waste products to ecochemicals: fifty years research of plant secondary metabolism. *Journal of Phytochemistry* **68**, 2831–2846 (2007).

24. Fraenkel, G. S. The Raison d'Etre of secondary plant substances: these odd chemicals arose as a means of protecting plants from insects and now guide insects to food. *Science* **129**, 1466–1470 (1959).

25. Hartmann, T. The lost origin of chemical ecology in the late 19th century. *Proceedings of the National Academy of Sciences* **105**, 4541–4546 (2008).

26. Matsuura, H. N. & Fett-Neto, A. G. Plant alkaloids: main features, toxicity, and mechanisms of action. *Plant Toxins* **2**, 1–15 (2015).

27. Chabaane, Y., Marques Arce, C., Glauser, G. & Benrey, B. Altered capsaicin levels in domesticated chili pepper varieties affect the interaction between a

generalist herbivore and its ectoparasitoid. *Journal of Pest Science* **95**, 735–747 (2022).

28. Gepner, J., Hall, L. & Sattelle, D. Insect acetylcholine receptors as a site of insecticide action. *Nature* **276**, 188–190 (1978); Steppuhn, A. et al. Nicotine's defensive function in nature. *PLOS Biology* **2**, e217 (2004).

29. Kim, Y.-S., Choi, Y.-E. & Sano, H. Plant vaccination: stimulation of defense system by caffeine production in planta. *Plant Signaling & Behavior* **5**, 489–493 (2010).

30. Nathanson, J. A., Hunnicutt, E. J., Kantham, L. & Scavone, C. Cocaine as a naturally occurring insecticide. *Proceedings of the National Academy of Sciences* **90**, 9645–9648 (1993).

31. Kirsch & Ogas, *Drug Hunters*.

32. Knudsmark Jessing, K., Duke, S. O. & Cedergreeen, N. Potential ecological roles of artemisinin produced by *Artemisia annua* L. *Journal of Chemical Ecology* **40**, 100 (2014).

33. Gershenzon, J. & Dudareva, N. The function of terpene natural products in the natural world. *Nature Chemical Biology* **3**, 408–414 (2007).

34. Griese, E. et al. Insect egg-killing: a new front on the evolutionary arms-race between brassicaceous plants and pierid butterflies. *New Phytologist* **230**, 341–353 (2021).

35. Durner, J., Shah, J. & Klessig, D. F. Salicylic acid and disease resistance in plants. *Trends in Plant Science* **2**, 266–274 (1997).

36. Moore, K. S. et al. Squalamine: an aminosterol antibiotic from the shark. *Proceedings of the National Academy of Sciences* **90**, 1354–1358 (1993).

37. Povey, S., Cotter, S. C., Simpson, S. J., Lee, K. P. & Wilson, K. Can the protein costs of bacterial resistance be offset by altered feeding behaviour? *Journal of Animal Ecology* **78**, 437–446 (2009).

38. Miller, C. V. & Cotter, S. C. Resistance and tolerance: the role of nutrients on pathogen dynamics and infection outcomes in an insect host. *Journal of Animal Ecology* **87**, 500–510 (2018).

39. Adamo, S. A., Bartlett, A., Le, J., Spencer, N. & Sullivan, K. Illness-induced anorexia may reduce trade-offs between digestion and immune function. *Animal Behaviour* **79**, 3–10 (2010).

40. Lee, K. P., Cory, J. S., Wilson, K., Raubenheimer, D. & Simpson, S. J. Flexible diet choice offsets protein costs of pathogen resistance in a caterpillar. *Proceedings of the Royal Society of London Series B: Biological Sciences* **273**, 823–829 (2006).

41. Kyriazakis, I., Anderson, D., Oldham, J., Coop, R. & Jackson, F. Long-term subclinical infection with *Trichostrongylus colubriformis*: effects on food intake, diet selection and performance of growing lambs. *Veterinary Parasitology* **61**, 297–313 (1996).

42. Lee et al., Flexible diet choice.

43. Povey et al., Can the protein costs of bacterial resistance be offset?

44. Cotter, S. C. et al. Diet modulates the relationship between immune gene expression and functional immune responses. *Journal of Insect Biochemistry and Molecular Biology* **109**, 128–141 (2019).

45. Wilson, K. et al. Osmolality as a novel mechanism explaining diet effects on the outcome of infection with a blood parasite. *Current Biology* **30**, 2459–2467.e3 (2020).

46. Stolz, J. L. *Medicine from Cave Dwellers to Millennials* (Telemachus Press, 2018).

47. Cardenas, D. Let not thy food be confused with thy medicine: the Hippocratic misquotation. *e-SPEN Journal* **8**, e260–e262 (2013).

48. Stolz, *Medicine from Cave Dwellers to Millennials.*

49. Borzelleca, J. F. Paracelsus: herald of modern toxicology. *Toxicological Sciences* **53**, 2–4 (2000).

8. Living and Learning

1. Conversation by video call between M. S. Singer and the author, 5 January 2022.

2. Bernays, E. A. & Singer, M. S. Insect defences: taste alteration and endoparasites. *Nature* **436**, 476 (2005).

3. Zuk, M. *Dancing Cockatoos and the Dead Man Test: How Behavior Evolves and Why It Matters* (W. W. Norton, 2022).

4. Rothenbuhler, W. C. Behavior genetics of nest cleaning in honey bees. IV. Responses of F 1 and backcross generations to disease-killed brood. *American Zoologist* **4**, 111–123 (1964).

5. Dawkins, R. *The Selfish Gene* (Oxford University Press, 1976).

6. Bastock, M. A gene mutation which changes a behavior pattern. *Evolution* **10**, 421–439 (1956).

7. Zuk, *Dancing Cockatoos.*

8. Tallamy, D. W., Darlington, M. B., Pesek, J. D. & Powell, B. E. Copulatory courtship signals male genetic quality in cucumber beetles. *Proceedings of the Royal Society of London Series B: Biological Sciences* **270**, 77–82 (2003).

9. Snell-Rood, E. C. & Papaj, D. R. Patterns of phenotypic plasticity in common and rare environments: a study of host use and color learning in the cabbage white butterfly *Pieris rapae. American Naturalist* **173**, 615–631 (2009).

10. Hart, B. L. The evolution of herbal medicine: behavioural perspectives. *Animal Behaviour* **70**, 975–989 (2005).

11. Handal, P. J. Immediate acceptance of sodium salts by sodium deficient rats. *Psychonomic Science* **3**, 315–316 (1965).

12. Vitazkova, S. K., Long, E., Paul, A. & Glendinning, J. I. Mice suppress malaria infection by sampling a "bitter" chemotherapy agent. *Animal Behaviour* **61**, 887–894 (2001).

13. Koshimizu, K., Ohigashi, H. & Huffman, M. A. Use of *Vernonia amygdalina* by wild chimpanzee: possible roles of its bitter and related constituents. *Physiology & Behavior* **56**, 1209–1216 (1994).

14. Zuk, *Dancing Cockatoos.*

15. Howard, S. R., Avarguès-Weber, A., Garcia, J. E., Greentree, A. D. & Dyer, A. G. Symbolic representation of numerosity by honeybees (*Apis mellifera*): matching characters to small quantities. *Proceedings of the Royal Society B: Biological Sciences* **286**, 20190238 (2019).

16. Bielecki, J., Nielsen, S.K.D., Nachman, G. & Garm, A. Associative learning in the box jellyfish *Tripedalia cystophora. Current Biology* **33**, 4150–4159 (2023).

17. Green, K. F. & Garcia, J. Recuperation from illness: flavor enhancement for rats. *Science* **173**, 749–751 (1971).

18. Zahorik, D. M. Associative and non-associative factors in learned food preferences. In *Learning Mechanisms in Food Selection* (ed. L. M. Barker, M. R. Best, & M. Domjan) pp. 181–199 (Baylor University Press, 1977).

19. Provenza, F. *Nourishment: What Animals Can Teach Us About Discovering Our Nutritional Wisdom* (Chelsea Green Publishing, 2018).

20. Sahley, C., Gelperin, A. & Rudy, J. W. One-trial associative learning modifies food odor preferences of a terrestrial mollusc. *Proceedings of the National Academy of Sciences* **78**, 640–642 (1981); Lee, J. C. & Bernays, E. A. Food tastes and toxic effects: associative learning by the polyphagous grasshopper *Schistocerca americana* (Drury) (Orthoptera: Acrididae). *Animal Behaviour* **39**, 163–173 (1990); Burghardt, G. M., Wilcoxon, H. C. & Czaplicki, J. A. Conditioning in garter snakes: aversion to palatable prey induced by delayed illness. *Animal Learning & Behavior* **1**, 317–320 (1973); MacKay, B. Conditioned food aversion produced by toxicosis in Atlantic cod. *Behavioral Biology* **12**, 347–355 (1974); Nicolaus, L. K., Cassel, J. F., Carlson, R. B. & Gustavson, C. R. Taste-aversion conditioning of crows to control predation on eggs. *Science* **220**, 212–214 (1983).

21. Lin, J.-Y., Arthurs, J. & Reilly, S. Conditioned taste aversions: from poisons to pain to drugs of abuse. *Psychonomic Bulletin Review* **24**, 335–351 (2017).

22. Jacobsen, P. B. et al. Formation of food aversions in cancer patients receiving repeated infusions of chemotherapy. *Behaviour Research and Therapy* **31**, 739–748 (1993).

23. Rozin, P. The selection of foods by rats, humans, and other animals. In: *Advances in the Study of Behavior,* vol. 6, pp. 21–76 (Elsevier, 1976).

24. Rozin, P. Specific aversions as a component of specific hungers. *Journal of Comparative Physiological Psychology* **64**, 237–242 (1967).

25. Lin, Arthurs, & Reilly, Conditioned taste aversions.

26. Rozin, Selection of foods by rats, humans, and other animals.

27. Etscorn, F. & Stephens, R. Establishment of conditioned taste aversions with a 24-hour CS-US interval. *Physiological Psychology* **1**, 251–253 (1973).

28. Goodall, J. *Seeds of Hope: Wisdom and Wonder from the World of Plants* (Grand Central Publishing, 2014).

29. Vanina Egea, A., Hall, J. O., Miller, J., Spackman, C. & Villalba, J. J. Reduced neophobia: a potential mechanism explaining the emergence of self-medicative behavior in sheep. *Physiology & Behavior* **135**, 189–197 (2014); Lisonbee, L. D., Villalba, J. J., Provenza, F. D. & Hall, J. O. Tannins and self-medication: implications for sustainable parasite control in herbivores. *Behavioural Processes* **82**, 184–189 (2009).

30. Gustafsson, E., Saint Jalme, M., Bomsel, M.-C. & Krief, S. Food neophobia and social learning opportunities in great apes. *International Journal of Primatology* **35**, 1037–1071 (2014).

31. Sanga, U., Provenza, F. D. & Villalba, J. J. Transmission of self-medicative behaviour from mother to offspring in sheep. *Animal Behaviour* **82**, 219–227 (2011).

32. de Waal, F. *The Ape and the Sushi Master* (Basic Books, 2001).

33. Galef, B. G. & Whiskin, E. E. "Conformity" in Norway rats? *Animal Behaviour* **75**, 2035–2039 (2008).

34. Yoerg, S. I. Social feeding reverses learned flavor aversions in spotted hyenas (*Crocuta crocuta*). *Journal of Comparative Psychology* **105**, 185 (1991).

35. Thornton, A. Social learning about novel foods in young meerkats. *Animal Behaviour* **76**, 1411–1421 (2008).

36. Dubost, J.-M. et al. Interspecific medicinal knowledge and Mahout-Elephant interactions in Thongmyxay district, Laos. *Revue d'ethnoécologie* **22**, https://doi.org/10.4000/ethnoecologie.9705 (2022); Lee, P. C. & Moss, C. J. The social context for learning and behavioural development among wild African elephants. In: *Mammalian Social Learning: Comparative Ecological Perspectives* (ed. H. Box & H. Gibson) pp. 102–125 (Cambridge University Press, 1999).

37. Nicol, C. J. Development, direction, and damage limitation: social learning in domestic fowl. *Learning and Behavior* **32**, 72–81 (2004).

38. Masi, S. et al. Unusual feeding behavior in wild great apes, a window to understand origins of self-medication in humans: role of sociality and physiology on learning process. *Physiology & Behavior* **105**, 337–349 (2012).

39. Huffman, M. A. & Seifu, M. Observations on the illness and consumption of a possibly medicinal plant *Vernonia amygdalina* (Del.), by a wild chimpanzee in the Mahale Mountains National Park, Tanzania. *Primates* **30**, 51–63 (1989).

40. Huffman, M. A., Spiezio, C., Sgaravatti, A. & Leca, J.-B. Leaf swallowing behavior in chimpanzees (*Pan troglodytes*): biased learning and the emergence of group level cultural differences. *Animal Cognition* **13**, 871–880 (2010).

41. Barelli, C. & Huffman, M. A. Leaf swallowing and parasite expulsion in Khao Yai white-handed gibbons (*Hylobates lar*), the first report in an Asian ape species. *American Journal of Primatology* **79**, e22610 (2017); Huffman, M. A. Self-medication: passive prevention and active treatment. In: *Encyclopedia of Animal Behavior* (ed. J. C. Choe) vol. 2, pp. 696–702 (Elsevier, Academic Press, 2019); Su, H.-h., Su, Y.-c. & Huffman, M. A. Leaf swallowing and parasitic infection of the Chinese lesser civet *Viverricula indica* in northeastern Taiwan. *Zoological Studies* **52**, 1–8 (2013).

42. Huffman, M. A. & Wrangham, R. W. Diversity of medicinal plant use by chimpanzees in the wild. In: *Chimpanzee Cultures* (ed. R. W. Wrangham, W. C. McGrew, F.B.M. de Waal, & P. G. Heltne) pp. 129–148 (Harvard University Press, 1994).

43. Menzel, C., Fowler, A., Tennie, C. & Call, J. Leaf surface roughness elicits leaf swallowing behavior in captive chimpanzees (*Pan troglodytes*) and bonobos (*P. paniscus*), but not in gorillas (*Gorilla gorilla*) or orangutans (*Pongo abelii*). *International Journal for Primatology* **34**, 533–553 (2013).

44. Conversations between Michael A. Huffman and the author. By video call, September and October 2021, and in person in Japan, March 2023.

45. de Waal, *Ape and Sushi Master*.

9. Woolly Wisdom

1. Darwin, C. *On the Origin of Species by Means of Natural Selection: or the Preservation of Favoured Races in the Struggle for Life* (John Murray, 1859).

2. Wallace, A. R. On the tendency of varieties to depart indefinitely from the original type. *Journal of the Proceedings of the Linnean Society London, Zoology* **3**, 45–62 (1858).

3. Conversation by video call between F. Provenza and the author, 8 October 2021.

4. Villalba, J. J. & Provenza, F. D. Preference for wheat straw by lambs conditioned with intraruminal infusions of starch. *British Journal of Nutrition* **77**, 287–297 (1997).

5. Phy, T. S. & Provenza, F. D. Sheep fed grain prefer foods and solutions that attenuate acidosis. *Journal of Animal Science* **76**, 954–960 (1998).

6. Provenza, F. *Nourishment: What Animals Can Teach Us About Discovering Our Nutritional Wisdom* (Chelsea Green Publishing, 2018).

7. Conversation by video call between J. Villalba and the author, 4 November 2021.

8. Lisonbee, L. D., Villalba, J. J., Provenza, F. D. & Hall, J. O. Tannins and self-medication: implications for sustainable parasite control in herbivores. *Behavioural Processes* **82**, 184–189 (2009).

9. Juhnke, J., Miller, J., Hall, J. O., Provenza, F. & Villalba, J. J. Preference for condensed tannins by sheep in response to challenge infection with *Haemonchus contortus*. *Veterinary Parasitology* **188**, 104–114 (2012).

10. Amit, M. et al. Self-medication with tannin-rich browse in goats infected with gastro-intestinal nematodes. *Veterinary Parasitology* **198**, 305–311 (2013).

11. Martínez-Ortíz-de-Montellano, C. et al. Effect of a tropical tannin-rich plant *Lysiloma latisiliquum* on adult populations of *Haemonchus contortus* in sheep. *Veterinary Parasitology* **172**, 283–290 (2010).

12. Kyriazakis, I., Anderson, D., Oldham, J., Coop, R. & Jackson, F. Long-term subclinical infection with *Trichostrongylus colubriformis*: effects on food intake, diet selection and performance of growing lambs. *Veterinary Parasitology* **61**, 297–313 (1996).

13. Poli, C. H. et al. Self-selection of plant bioactive compounds by sheep in response to challenge infection with *Haemonchus contortus*. *Physiology & Behavior* **194**, 302–310 (2018).

14. Atwood, S. B., Provenza, F. D., Wiedmeier, R. D. & Banner, R. E. Influence of free-choice vs mixed-ration diets on food intake and performance of fattening calves. *Journal of Animal Science* **79**, 3034–3040 (2001).

15. Simpson, S. J. & Raubenheimer, D. Obesity: the protein leverage hypothesis. *Obesity Reviews* **6**, 133–142 (2005); Simpson, S. J., Batley, R. & Raubenheimer, D. Geometric analysis of macronutrient intake in humans: the power of protein? *Appetite* **41**, 123–140 (2003).

16. Sanga, U., Provenza, F. D. & Villalba, J. J. Transmission of self-medicative behaviour from mother to offspring in sheep. *Animal Behaviour* **82**, 219–227 (2011).

17. Distel, R. A., Arroquy, J. I., Lagrange, S. & Villalba, J. J. Designing diverse agricultural pastures for improving ruminant production systems. *Frontiers in Sustainable Food Systems* **4**, 596869 (2020).

18. Totty, V., Greenwood, S., Bryant, R. H. & Edwards, G. Nitrogen partitioning and milk production of dairy cows grazing simple and diverse pastures. *Journal of Dairy Science* **96**, 141–149 (2013); Jonker, A. et al. Methane and carbon dioxide emissions from lactating dairy cows grazing mature ryegrass/white clover or a diverse pasture comprising ryegrass, legumes and herbs. *Animal Production Science* **59**, 1063–1069 (2018).

19. Distel et al., Designing diverse agricultural pastures; Lagrange, S., Beauchemin, K. A., MacAdam, J. & Villalba, J. J. Grazing diverse combinations of tanniferous and non-tanniferous legumes: implications for beef cattle performance and environmental impact. *Science of the Total Environment* **746**, 140788 (2020).

10. Sticky Bee Business

1. Gallai, N., Salles, J.-M., Settele, J. & Vaissiere, B. E. Economic valuation of the vulnerability of world agriculture confronted with pollinator decline. *Ecological Economics* **68**, 810–821 (2009).

2. Aurell, D., Bruckner, S., Wilson, M., Steinhauer, N. & Williams, G. R. A national survey of managed honey bee colony losses in the USA: results from the Bee Informed Partnership for 2020–21 and 2021–22. *Journal of Apicultural Research* **63**, 1–14 (2024).

3. Goulson, D., Nicholls, E., Botías, C. & Rotheray, E. L. Bee declines driven by combined stress from parasites, pesticides, and lack of flowers. *Science* **347**, 1255957 (2015).

4. Vanengelsdorp, D. & Meixner, M. D. A historical review of managed honey bee populations in Europe and the United States and the factors that may affect them. *Journal of Invertebrate Pathology* **103**, S80–S95 (2010).

5. Brosi, B. J., Delaplane, K. S., Boots, M. & de Roode, J. C. Ecological and evolutionary approaches to managing honeybee disease. *Nature Ecology and Evolution* **1**, 1250–1262 (2017).

6. Evans, J. D. et al. Immune pathways and defence mechanisms in honey bees *Apis mellifera*. *Insect Molecular Biology* **15**, 645–656 (2006).

7. Cremer, S., Armitage, S.A.O. & Schmid-Hempel, P. Social immunity. *Current Biology* **17**, R693–R702 (2007).

8. Starks, P. T., Blackie, C. A. & Seeley, T. D. Fever in honeybee colonies. *Naturwissenschaften* **87**, 229–231 (2000).

9. Simone-Finstrom, M. D. & Spivak, M. Increased resin collection after parasite challenge: a case of self-medication in honey bees? *PLOS ONE* **7**, e34601 (2012); Simone-Finstrom, M. & Spivak, M. Propolis and bee health: the natural history and significance of resin use by honey bees. *Apidologie* **41**, 295–311 (2010); Lemos, M. et al. *Baccharis dracunculifolia*, the main botanical source of Brazilian green propolis, displays antiulcer activity. *Journal of Pharmacy and Pharmacology* **59**, 603–608 (2007).

10. Conversation by video call between M. Spivak and the author, 6 October 2021.

11. Gekker, G., Hu, S., Spivak, M., Lokensgard, J. R. & Peterson, P. K. Anti-HIV-1 activity of propolis in CD4+ lymphocyte and microglial cell cultures. *Journal of Ethnopharmacology* **102**, 158–163 (2005).

12. Gekker et al., Anti-HIV-1 activity of propolis.

13. Hoyt, M. *The World of Bees*, p. 94 (Coward McCann, 1965).

14. Simone-Finstrom & Spivak, Propolis and bee health.

15. Hoyt, *World of Bees*, p. 93.

16. American foulbrood, BeeAware (website), https://beeaware.org.au/archive-pest/american-foulbrood/#ad-image-0, accessed 3 April 2024.

17. Bastos, E.M.A., Simone, M., Jorge, D. M., Soares, A.E.E. & Spivak, M. In vitro study of the antimicrobial activity of Brazilian propolis against *Paenibacillus larvae*. *Journal of Invertebrate Pathology* **97**, 273–281 (2008).

18. Simone, M., Evans, J. D. & Spivak, M. Resin collection and social immunity in honey bees. *Evolution* **63**, 3016–3022 (2009).

19. Simone-Finstrom & Spivak, Increased resin collection after parasite challenge.

20. Conversation by video call between M. Simone-Finstrom and the author, 19 November 2021.

21. Pusceddu, M. et al. Resin foraging dynamics in *Varroa destructor*-infested hives: a case of medication of kin? *Insect Science* **26**, 297–310 (2019); Drescher, N., Klein, A.-M., Neumann, P., Yañez, O. & Leonhardt, S. D. Inside honeybee hives: impact of natural propolis on the ectoparasitic mite *Varroa destructor* and viruses. *Insects* **8**, 15 (2017).

22. Pusceddu, M. et al. Honeybees use propolis as a natural pesticide against their major ectoparasite. *Proceedings of the Royal Society B: Biological Sciences* **288**, 20212101 (2021).

23. Borba, R. S. & Spivak, M. Propolis envelope in *Apis mellifera* colonies supports honey bees against the pathogen, *Paenibacillus larvae*. *Scientific Reports* **7**, 11429 (2017).

24. Dalenberg, H., Maes, P., Mott, B., Anderson, K. E. & Spivak, M. Propolis envelope promotes beneficial bacteria in the honey bee (*Apis mellifera*) mouthpart microbiome. *Insects* **11**, 453 (2020).

25. Corby-Harris, V. et al. *Parasaccharibacter apium*, gen. nov., sp nov., improves honey bee (Hymenoptera: Apidae) resistance to *Nosema*. *Journal of Economic Entomology* **109**, 537–543 (2016).

26. Endo, A. & Salminen, S. Honeybees and beehives are rich sources for fructophilic lactic acid bacteria. *Systematic and Applied Microbiology* **36**, 444–448 (2013).

27. Edwards, C., Haag, K., Collins, M., Hutson, R. & Huang, Y. *Lactobacillus kunkeei* sp. nov.: a spoilage organism associated with grape juice fermentations. *Journal of Applied Microbiology* **84**, 698–702 (1998).

28. Saelao, P., Borba, R. S., Ricigliano, V., Spivak, M. & Simone-Finstrom, M. Honeybee microbiome is stabilized in the presence of propolis. *Biology Letters* **16**, 20200003 (2020).

29. Simone-Finstrom & Spivak, Propolis and bee health.

30. Evans, J. D. Diverse origins of tetracycline resistance in the honey bee bacterial pathogen *Paenibacillus larvae*. *Journal of Invertebrate Pathology* **83**, 46–50 (2003).

31. Seeley, T. & Morse, R. The nest of the honey bee (*Apis mellifera* L.). *Insectes Sociaux* **23**, 495–512 (1976).

32. Nicodemo, D., De Jong, D., Couto, R. & Malheiros, B. Honey bee lines selected for high propolis production also have superior hygienic behavior and increased honey and pollen stores. *Genetics and Molecular Research* **12**, 6931–6938 (2013).

11. Dogs Are Dogs

1. Habib, R. & Becker, K. S. *The Forever Dog* (Harper Wave, 2021).

2. Krishnamani, R. & Mahaney, W. C. Geophagy among primates: adaptive significance and ecological consequences. *Animal Behaviour* **59**, 899–915 (2000).

3. Danford, D. E. Pica and nutrition. *Annual Review of Nutrition* **2**, 303–322 (1982).

4. Carretero, M. I. Clay minerals and their beneficial effects upon human health: a review. *Applied Clay Science* **21**, 155–163 (2002).

5. Gilardi, J. D., Duffey, S. S., Munn, C. A. & Tell, L. A. Biochemical functions of geophagy in parrots: detoxification of dietary toxins and cytoprotective effects. *Journal of Chemical Ecology* **25**, 897–922 (1999).

6. Houston, D., Gilardi, J. & Hall, A. Soil consumption by elephants might help to minimize the toxic effects of plant secondary compounds in forest browse. *Mammal Review* **31**, 249–254 (2001).

7. Krishnamani & Mahaney, Geophagy among primates; Pebsworth, P. A., Huffman, M. A., Lambert, J. E. & Young, S. L. Geophagy among nonhuman primates: a systematic review of current knowledge and suggestions for future directions. *American Journal of Physical Anthropology* **168**, 164–194 (2019).

8. Knezevich, M. Geophagy as a therapeutic mediator of endoparasitism in a free-ranging group of rhesus macaques (*Macaca mulatta*). *American Journal of Primatology* **44**, 71–82 (1998).

9. Zhou, D. et al. Soil causes gut microbiota to flourish and total serum IgE levels to decrease in mice. *Environmental Microbiology* **24**, 3898–3911 (2022).

10. Sun, X. et al. Harnessing soil biodiversity to promote human health in cities. *npj Urban Sustainability* **3**, 5 (2023); Blum, W. E., Zechmeister-Boltenstern, S. & Keiblinger, K. M. Does soil contribute to the human gut microbiome? *Microorganisms* **7**, 287 (2019).

11. Harris, E. V., de Roode, J. C. & Gerardo, N. M. Diet–microbiome–disease: investigating diet's influence on infectious disease resistance through alteration of the gut microbiome. *PLOS Pathogens* **15**, e1007891 (2019).

12. Bakken, J. S. et al. Treating *Clostridium difficile* infection with fecal microbiota transplantation. *Clinical Gastroenterology and Hepatology* **9**, 1044–1049 (2011); Fuentes, S. et al. Reset of a critically disturbed microbial ecosystem: faecal transplant in recurrent *Clostridium difficile* infection. *ISME Journal* **8**, 1621–1633 (2014); Kassam, Z., Lee, C. H., Yuan, Y. H. & Hunt, R. H. Fecal microbiota transplantation for *Clostridium difficile* infection: systematic review and meta-analysis. *American Journal of Gastroenterology* **108**, 500–508 (2013); Chang, J. Y. et al. Decreased diversity of the fecal microbiome in recurrent *Clostridium difficile*–associated diarrhea. *Journal of Infectious Diseases* **197**, 435–438 (2008).

13. Aristotle. *Aristotle's History of Animals in Ten Books*. Trans. Richard Creswell, Project Gutenberg EBook 59058 (George Bell & Sons, 1887).

14. Kandwal, M. K. & Sharma, M. *Cynodon dactylon* (L.) Pers.: a self-treatment grass for dogs. *Current Science* **101**, 619–621 (2011).

15. Hart, B. L. Why do dogs and cats eat grass? *Veterinary Medicine* **103**, 648–649 (2008); Sueda, K.L.C., Hart, B. L. & Cliff, K. D. Characterisation of plant eating in dogs. *Applied Animal Behaviour Science* **111**, 120–132 (2008).

16. Bjone, S., Brown, W. & Price, I. Grass eating patterns in the domestic dog, *Canis familiaris. Recent Advances in Animal Nutrition in Australia* **16**, 45–49 (2007).

17. McKenzie, S. J., Brown, W. Y. & Price, I. R. Reduction in grass eating behaviours in the domestic dog, *Canis familiaris*, in response to a mild gastrointestinal disturbance. *Applied Animal Behaviour Science* **123**, 51–55 (2010).

18. Bjone, Brown & Price, Grass eating patterns in the domestic dog.

19. Shipman, P. *Our Oldest Companions: The Story of the First Dogs* (Harvard University Press, 2021).

20. Murie, A. *The Wolves of Mount McKinley* (US Government Printing Office, 1944); Kuyt, E. *Feeding Ecology of Wolves on Barren-Ground Caribou Range in the Northwest Territories* (master's thesis, University of Saskatchewan, 1969).

21. Franck, A. R. & Farid, A. Many species of the Carnivora consume grass and other fibrous plant tissues. *Belgian Journal of Zoology* **150**, 1–70 (2020).

22. Franck & Farid, Many species of the Carnivora consume grass.

23. Toweill, D. E. & Maser, C. Food of cougars in the Cascade Range of Oregon. *Great Basin Naturalist* **45**, 77–80 (1985); Rollings, C. T. Habits, foods and parasites of the bobcat in Minnesota. *Journal of Wildlife Management* **9**, 131–145 (1945); Su, H.-h., Su, Y.-c. & Huffman, M. A. Leaf swallowing and parasitic infection of the Chinese lesser civet *Viverricula indica* in northeastern Taiwan. *Zoological Studies* **52**, 1–8 (2013); Huffman, M. A. Current evidence for self-medication in primates: a multidisciplinary perspective. *Yearbook of Physical Anthropology* **40**, 171–200 (1997).

24. Su, Su & Huffman, Leaf swallowing and parasitic infection of the Chinese lesser civet; Robinette, W. L., Gashwiler, J. S. & Morris, O. W. Food habits of the cougar in Utah and Nevada. *Journal of Wildlife Management* **23**, 261–273 (1959); Boukheroufa, M., Sakraoui, F., Belbel, F. & Sakraoui, R. Winter diet of the common genet, *Genetta genetta* (Carnivora, Viverridae), and the African golden wolf, *Canis anthus* (Carnivora, Canidae), in altitudinal locality of the Edough Forest (northeastern Algeria). *Zoodiversity* **54**, 67–74 (2020).

25. Huffman, M. & Caton, J. Self-induced increase of gut motility and the control of parasitic infections in wild chimpanzees. *International Journal of Primatology* **22**, 329–346 (2001).

26. Hart, B. L., Hart, L. A., Thigpen, A. P. & Willits, N. H. Characteristics of plant eating in domestic cats. *Animals* **11**, 1853 (2021).

27. Habib & Becker, *Forever Dog*.

28. van Asseldonk, T., Kleijer, G. & Lans, C. Ethnoveterinary herb use in the Netherlands: between ethnobotany and zoopharmacognosy. In: *Proceedings of the Conference: BASEL VET Preconference GA*, Basel, Switzerland, 1–6 (2017).

29. Schwartz, S. *Psychoactive Herbs in Veterinary Behavior Medicine* (Blackwell, 2005).

30. Ingraham, C. *Animal Self-Medication: How Animals Heal Themselves Using Essential Oils, Herbs and Minerals* (Ingraham Trading Ltd., 2018).

31. Ingraham, *Animal Self-Medication*.

32. Conversation by video call between C. Ingraham and the author, 13 February 2023.

33. Carretero, M. I. Clay minerals and their beneficial effects upon human health: a review. *Applied Clay Science* **21**, 155–163 (2002).

34. Carretero, Clay minerals and their beneficial effects.

35. Haydel, S. E., Remenih, C. M. & Williams, L. B. Broad-spectrum *in vitro* antibacterial activities of clay minerals against antibiotic-susceptible and antibiotic-resistant bacterial pathogens. *Journal of Antimicrobial Chemotherapy* **61**, 353–361 (2008).

36. Williams, L., Holland, M., Eberl, D., Brunet, T. & Brunet de Courrsou, L. Killer clays! natural antibacterial clay minerals. *Mineralogical Society Bulletin* **139**, 3–8 (2004).

37. Ezeorba, T.P.C. et al. Potentials for health and therapeutic benefits of garlic essential oils: recent findings and future prospects. *Pharmacological Research—Modern Chinese Medicine* **3**, 100075 (2022).

38. Xu, J.-G., Liu, T., Hu, Q.-P. & Cao, X.-M. Chemical composition, antibacterial properties and mechanism of action of essential oil from clove buds against *Staphylococcus aureus*. *Molecules* **21**, 1194 (2016).

39. Feng, J. et al. Identification of essential oils with strong activity against stationary phase *Borrelia burgdorferi*. *Antibiotics* **7**, 89 (2018); Feng, J. et al. Selective essential oils from spice or culinary herbs have high activity against stationary phase and biofilm *Borrelia burgdorferi*. *Frontiers in Medicine*, **169** (2017).

40. Conversation by video call between C. Hillier and the author, 13 April 2023.

12. Elephant Educators

1. Stolz, J. L. *Medicine from Cave Dwellers to Millennials* (Telemachus Press, 2018).

2. Kirsch, D. R. & Ogas, O. *The Drug Hunters* (Arcade Publishing, 2018).

3. Harvey, A. L. Natural products in drug discovery. *Drug Discovery Today* **13**, 894–901 (2008); Newman, D. J. & Cragg, G. M. Natural products as sources of new

drugs over the nearly four decades from 01/1981 to 09/2019. *Journal of Natural Products* **83**, 770–803 (2020).

4. Mandal, V., Gopal, V. & Mandal, S. C. An inside to the better understanding of the ethnobotanical route to drug discovery—the need of the hour. *Natural Product Communications* **7**, 1551–1554 (2012).

5. Quave, C. L. *The Plant Hunter: A Scientist's Quest for Nature's Next Medicines* (Viking, 2021).

6. Mandal, Gopal & Mandal, An inside to the better understanding.

7. Agosta, W. *Bombardier Beetles and Fever Trees: A Close-Up Look at Chemical Warfare and Signals in Animals and Plants* (Addison-Wesley, 1996).

8. Bendana, C. How a failed eczema treatment triggered an interest in traditional medicine. *Nature*, https://doi.org/10.1038/d41586-023-00168-0 (2023).

9. Huffman, M. A. Folklore, animal self-medication, and phytotherapy—something old, something new, something borrowed, some things true. *Planta Medica* **88**, 187–199 (2022).

10. Huffman, Folklore, animal self-medication, and phytotherapy.

11. Huffman, Folklore, animal self-medication, and phytotherapy.

12. Rockwell, D. *Giving Voice to Bear: North American Indian Myths, Rituals, and Images of the Bear* (Roberts Reinhart, 1991); Storl, W. D. *Bear: Myth, Animal, Icon* (North Atlantic Books, 2018).

13. Ma, H. et al. The genus *Epimedium*: an ethnopharmacological and phytochemical review. *Journal of Ethnopharmacology* **134**, 519–541 (2011).

14. Chen, M., Hao, J., Yang, Q. & Li, G. Effects of icariin on reproductive functions in male rats. *Molecules* **19**, 9502–9514 (2014); Liu, W. J. et al. Effects of icariin on erectile function and expression of nitric oxide synthase isoforms in castrated rats. *Asian Journal of Andrology* **7**, 381–388 (2005).

15. Ma et al., Genus *Epimedium*.

16. Ma et al., Genus *Epimedium*.

17. Riesenberg, S. H. Magic and medicine in Ponape. *Southwestern Journal of Anthropology* **4**, 406–429 (1948).

18. Huffman, Folklore, animal self-medication, and phytotherapy; Pope, H. G. *Tabernanthe iboga*: an African narcotic plant of social importance. *Economic Botany* **23**, 174–184 (1969).

19. Dubost, J.-M. et al. Zootherapeutic uses of animals excreta: the case of elephant dung and urine use in Sayaboury province, Laos. *Journal of Ethnobiology and Ethnomedicine* **17**, 62 (2021).

20. Greene, A. M., Panyadee, P., Inta, A. & Huffman, M. A. Asian elephant self-medication as a source of ethnoveterinary knowledge among Karen mahouts in northern Thailand. *Journal of Ethnopharmacology* **259**, 112823 (2020).

21. Khan, T., Khan, I. A., Rehman, A., Ali, S. & Ali, H. Zoopharmacognosy and epigenetic behavior of mountain wildlife towards Berberis species. *Life Science Journal* **11**, 259–263 (2014).

22. Gradé, J., Tabuti, J. R. & Van Damme, P. Four footed pharmacists: indications of self-medicating livestock in Karamoja, Uganda. *Economic Botany* **63**, 29–42 (2009).

23. Dubost, J.-M. et al. From plant selection by elephants to human and veterinary pharmacopeia of mahouts in Laos. *Journal of Ethnopharmacology* **244**, 112157 (2019).

24. Greene et al., Asian elephant self-medication.

25. Dubost, J.-M. et al. Interspecific medicinal knowledge and Mahout-Elephant interactions in Thongmyxay district, Laos. *Revue d'ethnoécologie* **22**, https://doi.org/10.4000/ethnoecologie.9705 (2022).

26. Gradé, Tabuti & Van Damme, Four footed pharmacists.

13. Cats and Catnip

1. Kirsch, D. R. & Ogas, O. *The Drug Hunters* (Arcade Publishing, 2018).

2. Jeffreys, D. *Aspirin: The Remarkable Story of a Wonder Drug* (Bloomsbury, 2004).

3. Quave, C. L. *The Plant Hunter: A Scientist's Quest for Nature's Next Medicines* (Viking, 2021).

4. Conversation between C. Quave and the author, Waller's Coffee Shop, Decatur, GA, 1 December 2021.

5. Forbey, J. S. et al. Exploitation of secondary metabolites by animals: a response to homeostatic challenges. *Integrative and Comparative Biology* **49**, 314–328 (2009); Raubenheimer, D. & Simpson, S. J. Nutritional PharmEcology: doses, nutrients, toxins, and medicines. *Integrative and Comparative Biology* **49**, 329–337 (2009).

6. Krief, S., Martin, M. T., Grellier, P., Kasenene, J. & Sevenet, T. Novel antimalarial compounds isolated in a survey of self-medicative behavior of wild chimpanzees in Uganda. *Antimicrobial Agents and Chemotherapy* **48**, 3196–3199 (2004).

7. Huffman, M. A. Chimpanzee self-medication: a historical perspective of the key findings. In: *Mahale Chimpanzees: 50 years of research* (ed. M. Nakamura, K. Hosaka, N. Itoh, & K. Zamma) pp. 340–353 (Cambridge University Press, 2015).

8. Conversation by video call between U. Maloueki and the author, 21 September 2022.

9. Spielman, A. & D'Antonio, M. *Mosquito: The Story of Man's Deadliest Foe* (Faber and Faber, 2001).

10. Conversation by video call between R. Uenoyama, M. Miyazaki, and the author, 22 August 2022.

11. Yoshitoshi, T. Accession number: 11.37780 (H. Enshûya, 31859) (William Sturgis Bigelow Collection).

12. Melo, N. et al. The irritant receptor TRPA1 mediates the mosquito repellent effect of catnip. *Current Biology* **31**, 1988–1994, e1985 (2021).

13. Maier, U. Agricultural activities and land use in a Neolithic village around 3900 BC: Hornstaad Hörnle IA, Lake Constance, Germany. *Vegetation History and Archaeobotany* **8**, 87–94 (1999).

14. Sakan, T., Fujino, A., Murai, F., Butsugan, Y. & Suzui, A. On the structure of actinidine and matatabilactone, the effective components of *Actinidia polygama*. *Bulletin of the Chemical Society of Japan* **32**, 315–316 (1959); Sakan, T., Fujino, A., Murai, F., Suzui, A. & Butsugan, Y. The structure of matatabilactone. *Bulletin of the Chemical Society of Japan* **32**, 1154–1155 (1959); Meinwald, J. The degradation of nepetalactone1. *Journal of the American Chemical Society* **76**, 4571–4573 (1954); McElvain, S., Walters, P. M. & Bright, R. D. The constituents of the volatile oil of catnip. II. The neutral components. Nepetalic anhydride. *Journal of the American Chemical Society* **64**, 1828–1831 (1942); McElvain, S. M., Bright, R. D. & Johnson, P. R. The constituents of the volatile oil of catnip. I. Nepetalic acid, nepetalactone and related compounds. *Journal of the American Chemical Society* **63**, 1558–1563 (1941).

15. Eisner, T. Catnip: its raison d'etre. *Science* **146**, 1318–1320 (1964).

16. Uenoyama, R. et al. The characteristic response of domestic cats to plant iridoids allows them to gain chemical defense against mosquitoes. *Science Advances* **7**, eabd9135 (2021).

17. Uenoyama et al., Characteristic response of domestic cats.

18. Uenoyama et al., Characteristic response of domestic cats.

19. Uenoyama, R. et al. Domestic cat damage to plant leaves containing iridoids enhances chemical repellency to pests. *iScience* **25**, 104455 (2022).

20. Koch, H. et al. Host and gut microbiome modulate the antiparasitic activity of nectar metabolites in a bumblebee pollinator. *Philosophical Transactions of the Royal Society B: Biological Sciences* **377**, 20210162 (2022).

21. Brütsch, T., Jaffuel, G., Vallat, A., Turlings, T. C. & Chapuisat, M. Wood ants produce a potent antimicrobial agent by applying formic acid on tree-collected resin. *Ecology and Evolution* **7**, 2249–2254 (2017).

22. Todd, N. B. Inheritance of the catnip response in domestic cats. *Journal of Heredity* **53**, 54–56 (1962).

14. Plants and Pollinators

1. Erwin, D. H. et al. The Cambrian conundrum: early divergence and later ecological success in the early history of animals. *Science* **334**, 1091–1097 (2011).

2. Ceballos, G. et al. Accelerated modern human–induced species losses: entering the sixth mass extinction. *Science Advances* **1**, e1400253 (2015).

3. Dirzo, R. et al. Defaunation in the Anthropocene. *Science* **345**, 401–406 (2014).

4. World Wildlife Fund. *Living Planet Report 2020: Bending the Curve of Biodiversity Loss* (Gland, Switzerland, 2020).

5. Pacheco, P. et al. *Deforestation Fronts: Drivers and Responses in a Changing World* (WWF, Gland, Switzerland, 2021).

6. Voigt, M. et al. Global demand for natural resources eliminated more than 100,000 Bornean orangutans. *Current Biology* **28**, 761–769, e765 (2018).

7. Dirzo, R. et al. Defaunation in the Anthropocene. *Science* **345**, 401–406 (2014).

8. Losey, J. E. & Vaughan, M. The economic value of ecological services provided by insects. *Bioscience* **56**, 311–323 (2006).

9. Goodall, J. *Seeds of Hope: Wisdom and Wonder from the World of Plants* (Grand Central Publishing, 2014).

10. Bromham, L. et al. Global predictors of language endangerment and the future of linguistic diversity. *Nature Ecology and Evolution* **6**, 163–173 (2022).

11. Cámara-Leret, R. & Bascompte, J. Language extinction triggers the loss of unique medicinal knowledge. *Proceedings of the National Academy of Sciences* **118**, e2103683118 (2021).

12. Natterson-Horowitz, B. & Bowers, K. *Zoobiquity* (Vintage, 2013).

13. Chapman, C. A. & Huffman, M. A. Refining thoughts about human/nonhuman differences. *Animal Sentience* **3**, 231 (2019).

14. Huffman, M. A. Folklore, animal self-medication, and phytotherapy—something old, something new, something borrowed, some things true. *Planta Medica* **88**, 187–199 (2022).

15. Folly, A. J., Koch, H., Farrell, I. W., Stevenson, P. C. & Brown, M. J. Agri-environment scheme nectar chemistry can suppress the social epidemiology of parasites in an important pollinator. *Proceedings of the Royal Society B: Biological Sciences* **288**, 20210363 (2021).

16. World Bank. Surface area: Netherlands, https://data.worldbank.org/indicator/AG.SRF.TOTL.K2?locations=NL&view=map (2018).

17. Kawahara, A. Y., Reeves, L. E., Barber, J. R. & Black, S. H. Opinion: eight simple actions that individuals can take to save insects from global declines. *Proceedings of the National Academy of Sciences* **118**, e2002547117 (2021).

18. Ur Rehman, F. et al. Importance of medicinal plants in human and plant pathology: a review. *International Journal of Pharmacy & Biomedical Research* **8**, 1–11 (2021).

19. Tausende gärten werden zu oasen für die biologische vielfalt (Thousands of gardens become oases for biodiversity), Butterfly Meadows (website), http://www.schmetterlingswiesen.de/PagesSw/ContentList.aspx?id=2089 (2020).

INDEX

A page number in *italics* refers to a figure.

GPSR Authorized Representative: Easy Access System Europe - Mustamäe tee 50, 10621 Tallinn, Estonia, gpsr.requests@easproject.com

www.ingramcontent.com/pod-product-compliance
Lightning Source LLC
Chambersburg PA
CBHW010109160926
51573CB00025B/326
9780691239354